Generalized SBC Process Algebra for Communication and Concurrency

-- The Structure-Behavior Coalescence Approach --

Third Edition

William S. Chao

Structure-Behavior Coalescence

Systems Architecture **=** **Systems Structure** **+** **Systems Behavior**

CONTENTS

6

PREFACE TO THE THIRD EDITION

After the second edition of this book has been published, we found that in addition to treating each "condition" in a conditional expression as a prefix, we can embed each condition in the prefix.

We decide to revise this book. In the third edition of this book, every condition is embedded into a prefix. All the theories of generalized SBC process algebra (G-SBC-PA), such as interleaving, restricted composition and structural composition are thereafter gorgeously developed.

Readers are encouraged to compare the first and second versions with this one. It is interesting to understand the author's rewrite of the twists and turns of this book.

PREFACE TO THE SECOND EDITION

In the first edition of this book, we do not treat "conditions" in conditional expressions as prefixes. Therefore, all the theories of generalized SBC process algebra (G-SBC-PA), such as interleaving, restricted composition, structural composition, observation equivalence and observation congruence, attempt to ignore conditional expressions.

After further research, we find that if we treat the "condition" as a prefix, all the theories of generalized SBC process algebra will be more complete and beautiful.

In the second edition of this book, the reader should like to read conditional expressions that are so elegantly handled in all the theories such as interleaving, restricted composition, structural composition, and so on.

PREFACE TO THE FIRST EDITION

Single-queue SBC process algebra (S-SBC-PA), multi-queue SBC process algebra (M-SBC-PA), infinite-queue SBC process algebra (I-SBC-PA) and generalized SBC process algebra (G-SBC-PA) possessing the characteristics of structure-behavior coalescence (SBC) and all evolved from CCS (Calculus of Communicating Systems). CCS is a general process algebra language for the study of concurrent systems.

In one way, three specialized SBC process algebras S-SBC-PA, M-SBC-PA and I-SBC-PA, less like CCS, are only applicable to systems architecture. In another way, generalized SBC process algebra, more like CCS, is a general process algebra language for the study of communication and concurrency.

We shall go through the details of generalized SBC process algebra in this book. Hopefully, readers will be able to identify the differences between CCS and G-SBC-PA.

ABOUT THE AUTHOR

Dr. William S. Chao is the CEO & founder of SBC Architecture International®. SBC (Structure-Behavior Coalescence) architecture is a systems architecture which demands the integration of systems structure and systems behavior of a system. SBC architecture applies to hardware architecture, software architecture, enterprise architecture, knowledge architecture and thinking architecture. The core theme of SBC architecture is: Architecture = Structure + Behavior.

William S. Chao received his bachelor degree (1976) in telecommunication engineering and master degree (1981) in information engineering, both from the National Chiao-Tung University, Taiwan. From 1976 till 1983, he worked as an engineer at Chung-Hwa Telecommunication Company, Taiwan.

William S. Chao received his master degree (1985) in information science and Ph.D. degree (1988) in information science, both from the University of Alabama at Birmingham, USA. From 1988 till 1991, he worked as a computer scientist at GE Research and Development Center, Schenectady, New York, USA.

Dr. William S. Chao has been teaching at National Sun Yat-Sen University, Taiwan since 1992 and now serves as the president of Association of Enterprise Architects, Taiwan Chapter. His research covers: systems architecture, hardware architecture, software architecture, enterprise architecture, knowledge architecture and thinking architecture.

PART I: WHAT IS PROCESS ALGEBRA?

18

Algebraic Approach to the Study of Concurrent Systems

Process algebras are a diverse family of related approaches to the study of concurrent systems [Berg87, Chao15d, Hoar85, Miln89, Miln99]. Their tools are algebraic languages for the high-level description of interactions, communications and synchronizations between a collection of independent agents or processes.

Process algebras also provide algebraic laws that allow process descriptions to be manipulated and analyzed, and permit formal reasoning about equivalences and observation congruence among processes.

Examples of Process Algebras

There are several leading algebraic approaches to modeling concurrent systems.

Communicating Sequential Processes (CSP) [Hoar85] was first described in a 1978 paper by C. A. R. Hoare.

Arthur John Robin Gorell Milner introduced the Calculus of Communicating Systems (CCS) [Miln89, Miln99] around 1980.

Algebra of Communicating Processes (ACP) [Berg87] was initially developed by Jan Bergstra and Jan Willem Klop in 1982.

Specialized SBC Process Algebras

Single-queue SBC process algebra (S-SBC-PA) [Chao15d, Chao15e, Chao15g], multi-queue SBC process algebra (M-SBC-PA) [Chao15d, Chao15f, Chao15h] and infinite-queue SBC process algebra (I-SBC-PA) [Chao15b, Chao15c, Chao15d] are the three specialized SBC process algebras.

Single-queue SBC process algebra, multi-queue SBC process algebra and infinite-queue SBC process algebra all evolved from CCS (Calculus of Communicating Systems) [Miln89, Miln99].

CCS is a general process algebra language for the study of concurrent systems. Unlike CCS, three specialized SBC process algebras are only applicable to systems architecture [Burd10, Maie09, Chao16b, Chao16c, Chao16d, Chao16e, Chao16f, Chao16g].

Generalized SBC Process Algebra

Generalized SBC process algebra (G-SBC-PA) also evolved from CCS (Calculus of Communicating Systems) [Miln89, Miln99].

CCS is a general process algebra language for the study of communication and concurrency. Like CCS, generalized SBC process algebra is also a general process algebra language for the study of communication and concurrency.

PART II: CHANNEL-BASED VALUE-PASSING INTERACTIONS

Channels

Channels are a model for agent communication. An agent may provide many channels, as shown in the figure below.

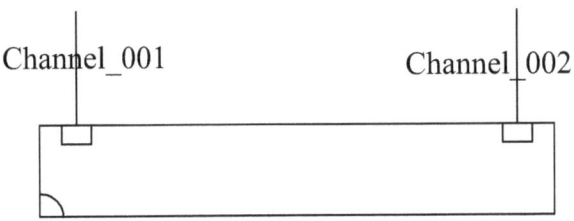

A channel may contain several input parameters (e.g. i_1, i_2) and output parameters (e.g. o_1, o_2).

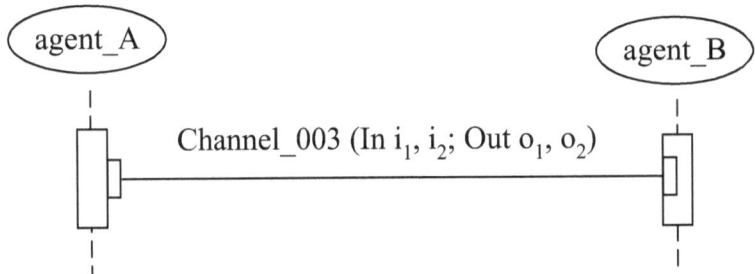

A channel formula is used to completely describe a channel. A channel formula includes a) channel name, b) input parameters (e.g. i_1, i_2, ..., i_m) and c) output parameters (e.g. o_1, o_2, ..., o_n).

Channel_Name (In i_1, i_2, ..., i_m; Out o_1, o_2, ..., o_n)

Channel-Based Interactions

A channel-based interaction represents an indivisible and instantaneous handshake or communication between two agents. The caller agent (either external environment's actor or component) interacts with the callee agent (component) through the channel interaction.

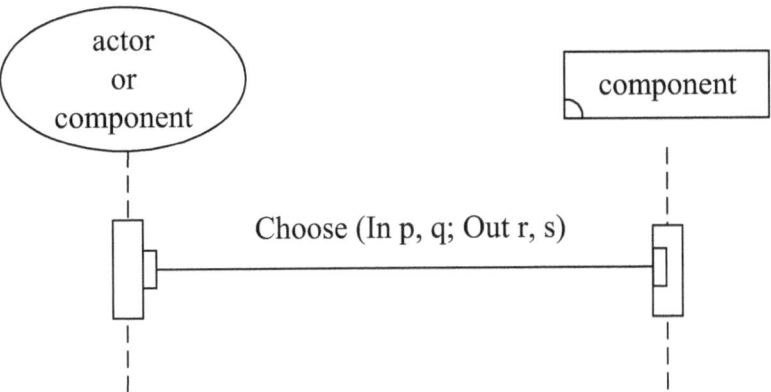

The caller agent owns the "calling port" of the interaction. In this case, the calling port is " $\overline{\text{Choose (In p, q; Out r, s)}}$ " and its conduct is to assist the caller agent to output a value to each of the "p" and "q" variables (of the "Choose" channel), and input a value from each of the "r" and "s" variables (of the "Choose" channel).

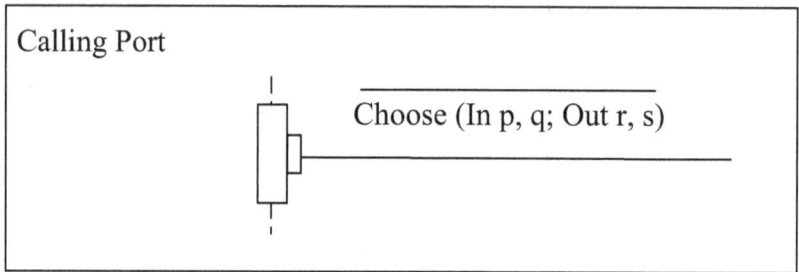

The caller agent together with the "calling port" is named the "calling action".

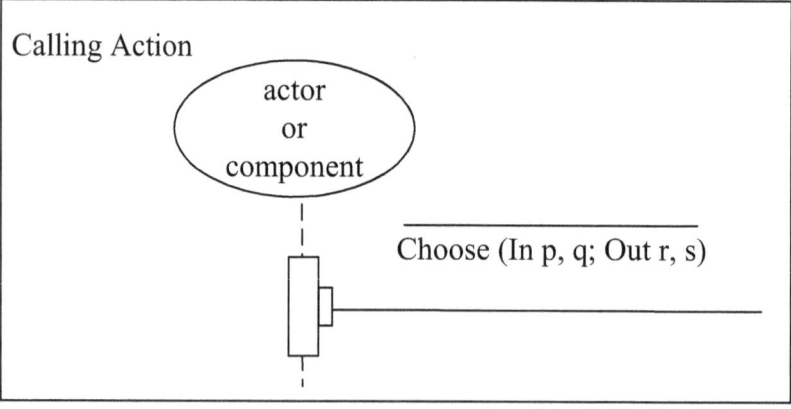

The callee agent owns the "called port" of the interaction. In this case, the called port is "Choose (In p, q; Out r, s)" and its conduct is to assist the callee agent to input a value from each of the "p" and "q" variables (of the "Choose" channel), and output a value to each of the "r" and "s" variables (of the "Choose" channel).

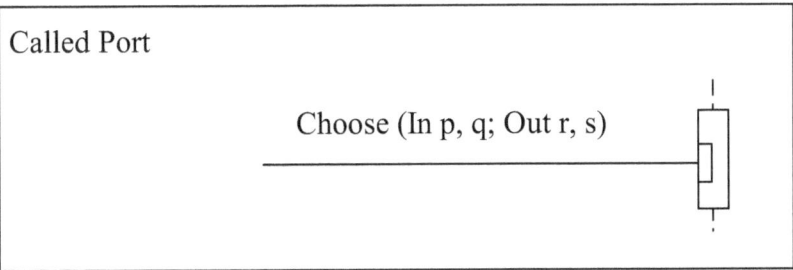

The callee agent together with the "called port" is named the "called action".

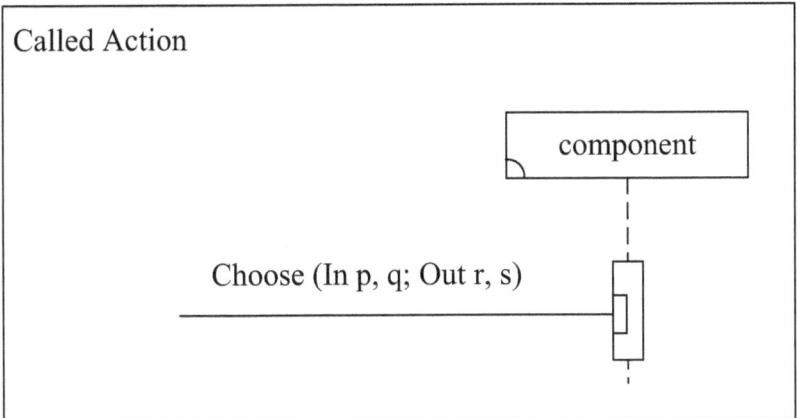

In order to simplify the channel-based interaction diagram, we will redraw it as follows.

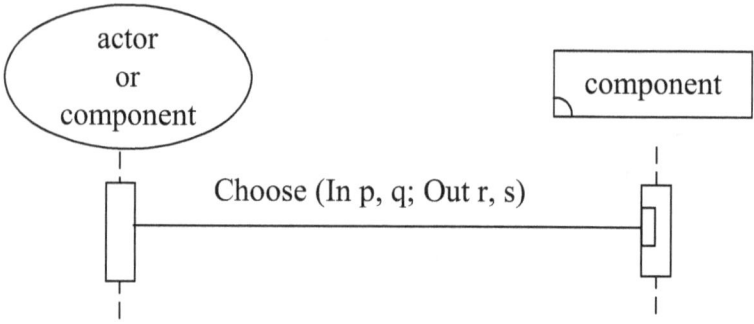

Or we can draw the channel-based interaction diagram as follows.

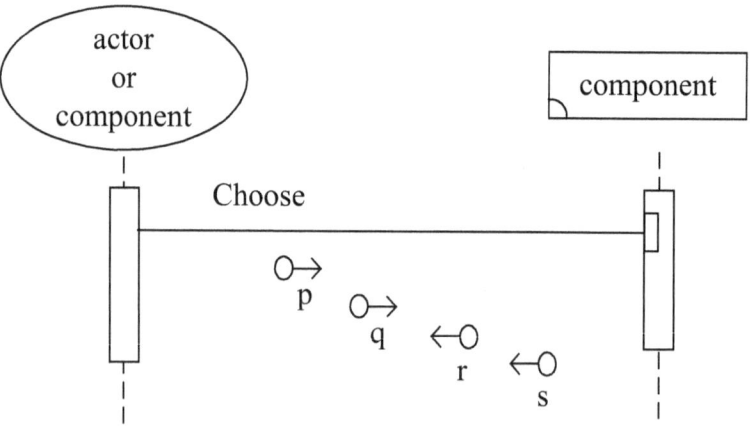

We use a channel-based internal interaction (i.e. λ) to represent their handshake or communication, if the caller agent and the callee agent are the same one.

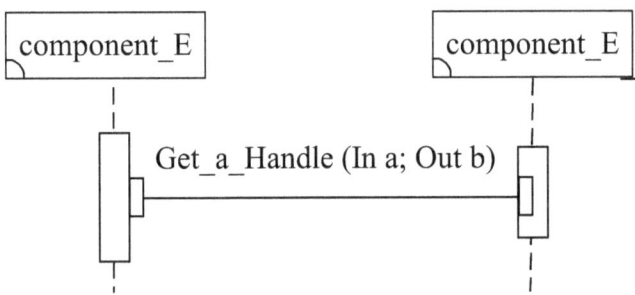

Also, we may redraw the channel-based internal interaction as follows.

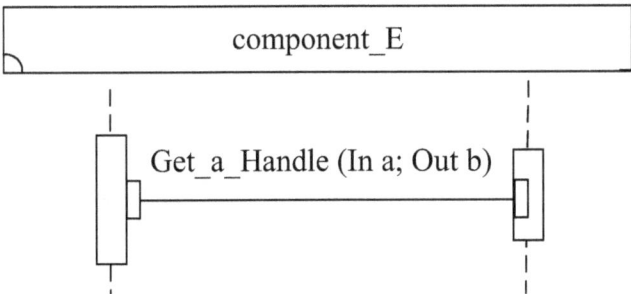

Formal Description of a Channel-Based Communication Port

We formally describe a channel-based communication port as a 2-tuple PORT = <calling_or_called, channel_formula>, where "calling_or_called" stands for a CALLING or CALLED port tag and "channel_formula" stands for a channel formula.

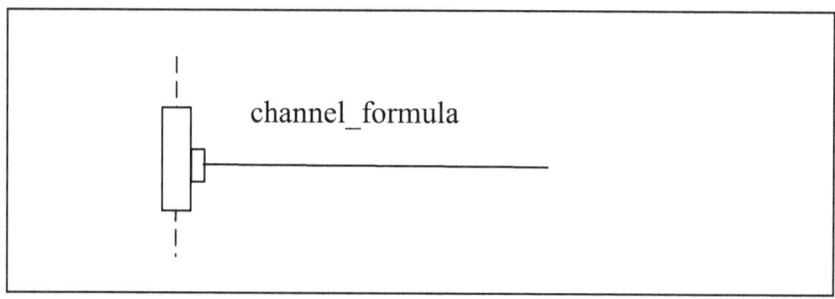

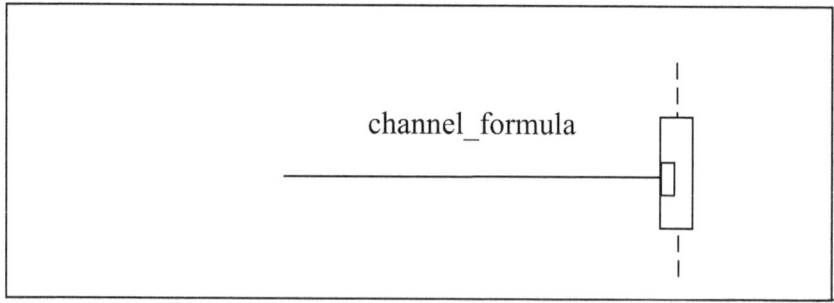

Formal Description of a Channel-Based Action

We formally describe a channel-based action as a 3-tuple ACTION = <agent, calling_or_called, channel_formula>, where "agent" stands for the name of a caller or callee agent, "calling_or_called" stands for a CALLING or CALLED action tag, and "channel_formula" stands for a channel formula.

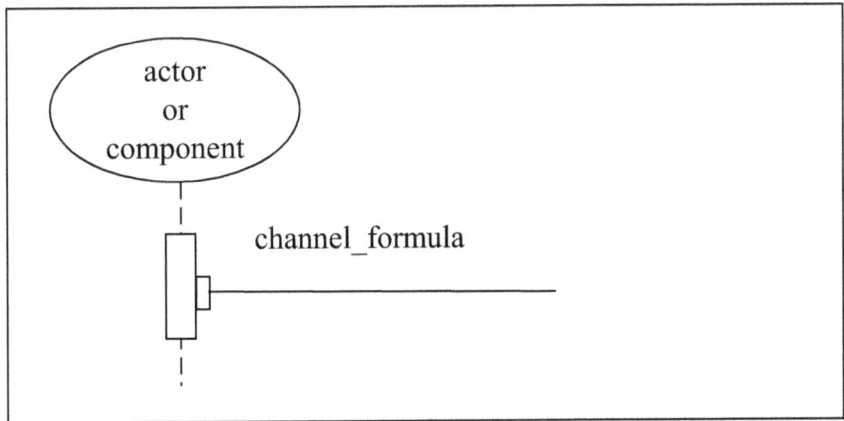

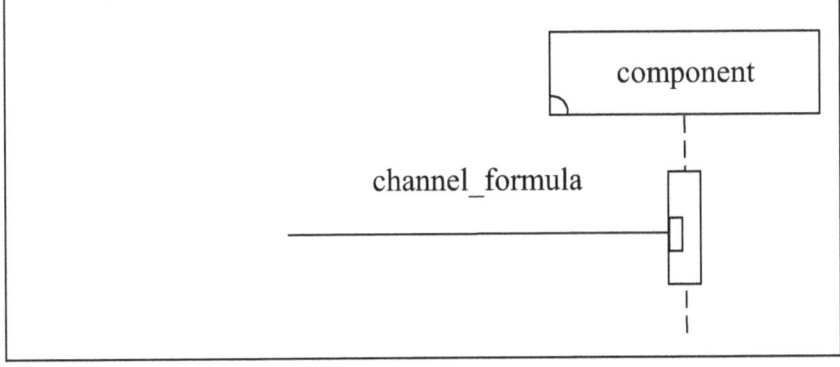

Formal Description of a Channel-Based Interaction

We formally describe a channel-based interaction as a 3-tuple INTERACTION = <caller_agent, channel_formula, callee_agent >, where "caller_agent" stands for the name of a caller agent, "channel_formula" stands for a channel formula and "callee_agent" stands for the name of a callee agent.

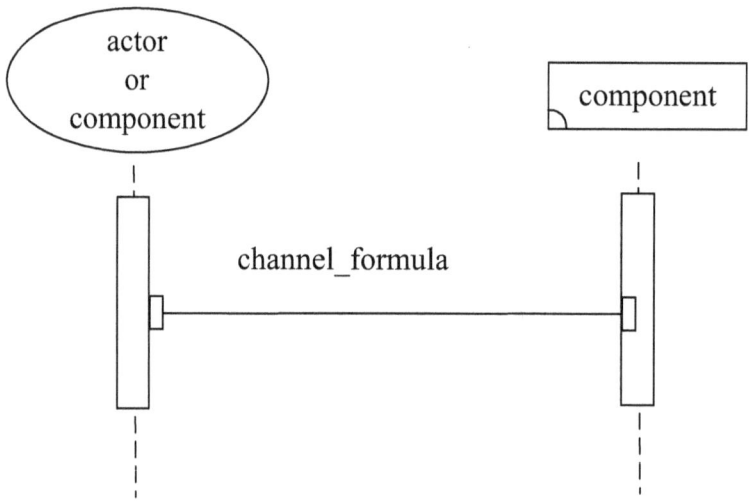

PART III: OPERATION-BASED VALUE-PASSING INTERACTIONS

Operations

An operation provided by each component represents a procedure, or method, or function of the component.

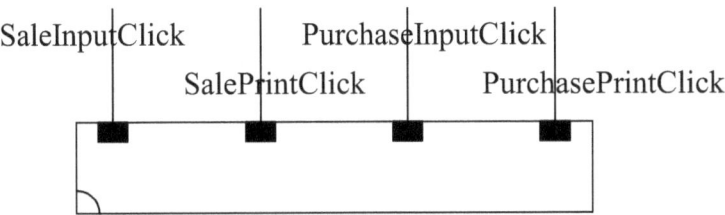

An operation may contain several input parameters (e.g. i_1, i_2) and output parameters (e.g. o_1, o_2).

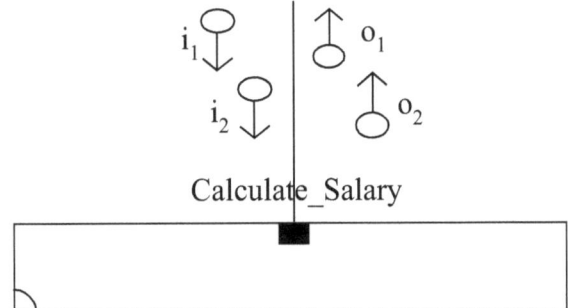

An operation formula is used to completely describe an operation. An operation formula includes a) operation name, b) input parameters (e.g. i_1, i_2, ..., i_m) and c) output parameters (e.g. o_1, o_2, ..., o_n).

$$\text{Operation\_Name (In } i_1, i_2, ..., i_m; \text{ Out } o_1, o_2, ..., o_n)$$

Operation-Based Interactions

An operation-based interaction represents an indivisible and instantaneous handshake or communication between two agents. The caller agent (either external environment's actor or component) communicates with the callee agent (component) through the operation call or return interaction (also named as operation call or reply message).

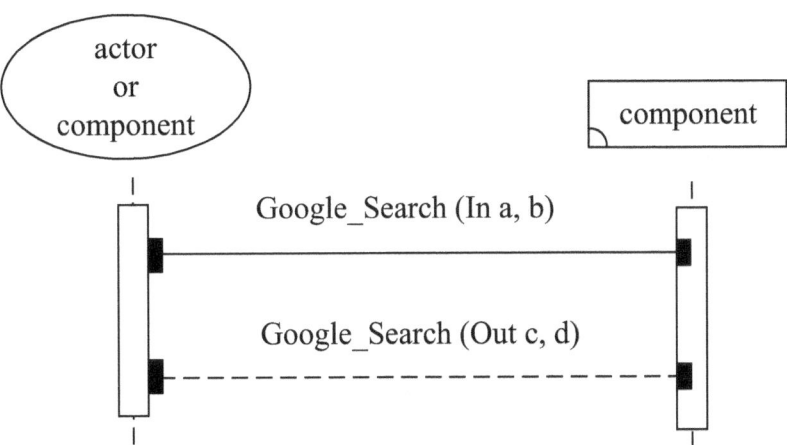

The caller agent owns the "calling port" of the interaction. In the operation call interaction (also known as operation call message) case, the calling port is " <u>Google_Search (In a, b)</u> " and its conduct is to assist the caller agent to output a value to each of the "a" and "b" variables (of the "Google_Search" operation).

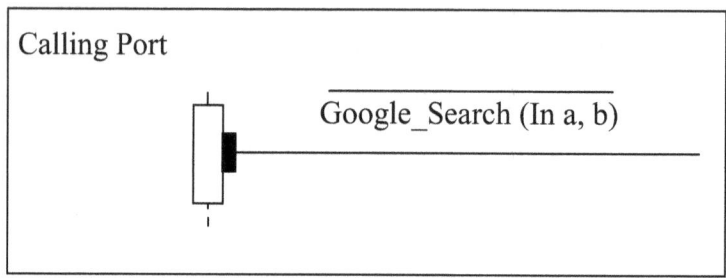

The caller agent together with the "calling port" is named the "calling action".

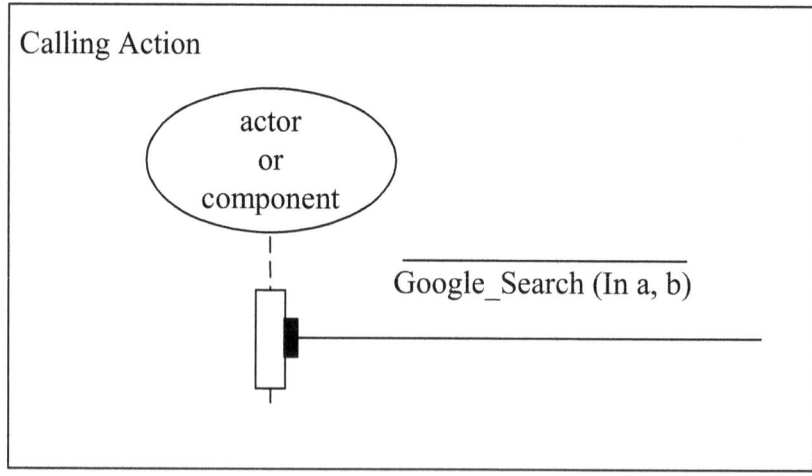

In the operation return interaction (also known as operation reply message) case, the calling port is " Google_Search (Out c, d) " and its conduct is to assist the caller agent to input a value from each of the "c" and "d" variables (of the "Google_Search" operation).

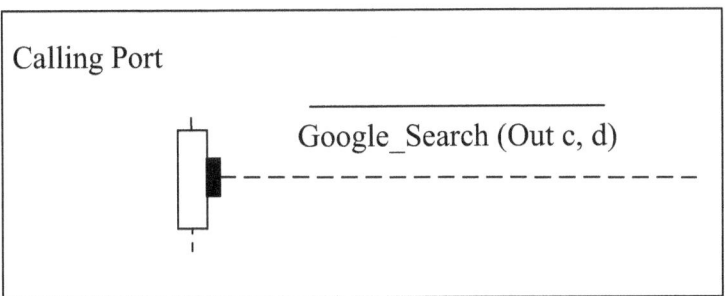

The caller agent together with the "calling port" is named the "calling action".

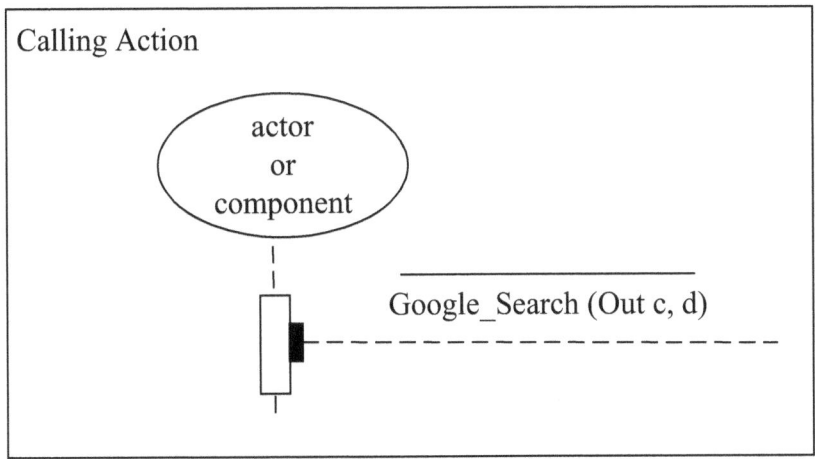

The callee agent owns the "called port" of the interaction. In the operation call interaction case, the called port is "Google_Search (In a, b)" and its conduct is to assist the callee agent to input a value from each of the "a" and "b" variables (of the "Google_Search" operation).

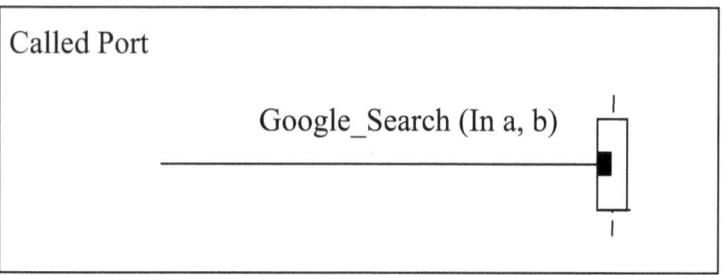

The callee agent together with the "called port" is named the "called action".

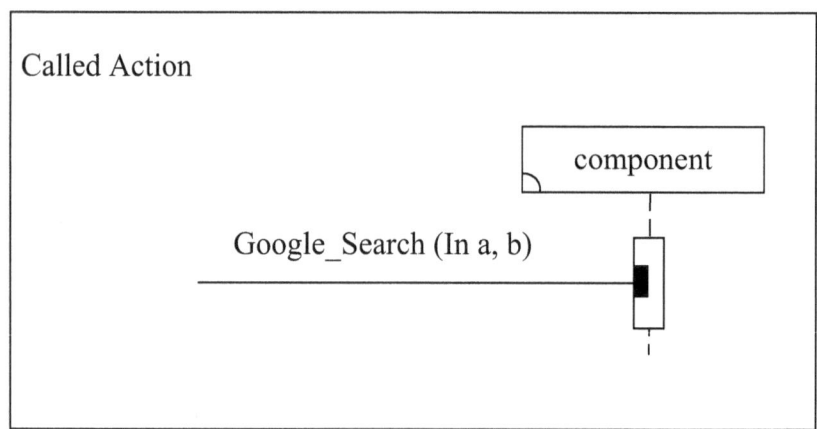

In the operation return interaction case, the called port is "Google_Search (Out c, d)" and its conduct is to assist the callee agent to output a value to each of the "c" and "d" variables (of the "Google_Search" operation).

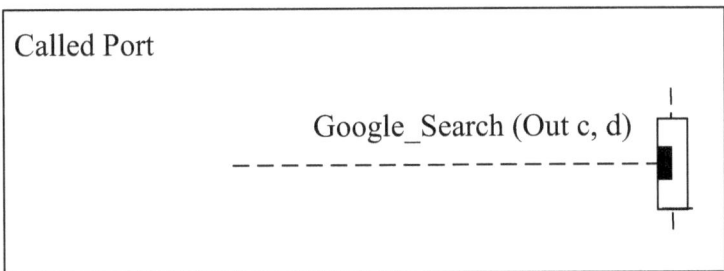

The callee agent together with the "called port" is named the "called action".

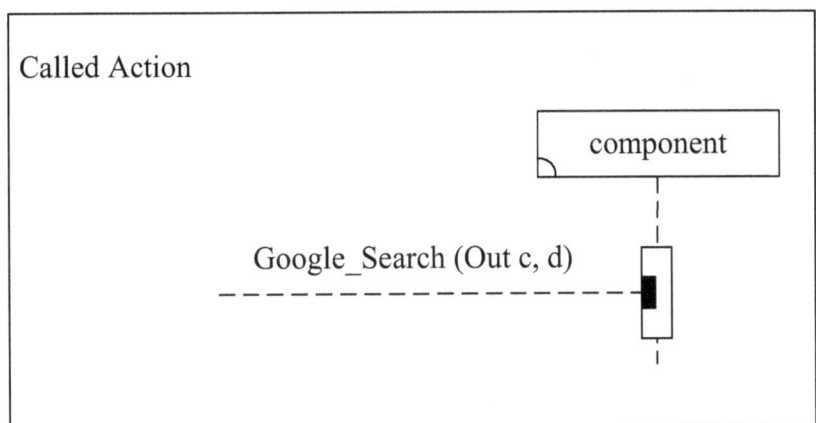

In order to simplify the operation-based interaction diagram, we will redraw it as follows.

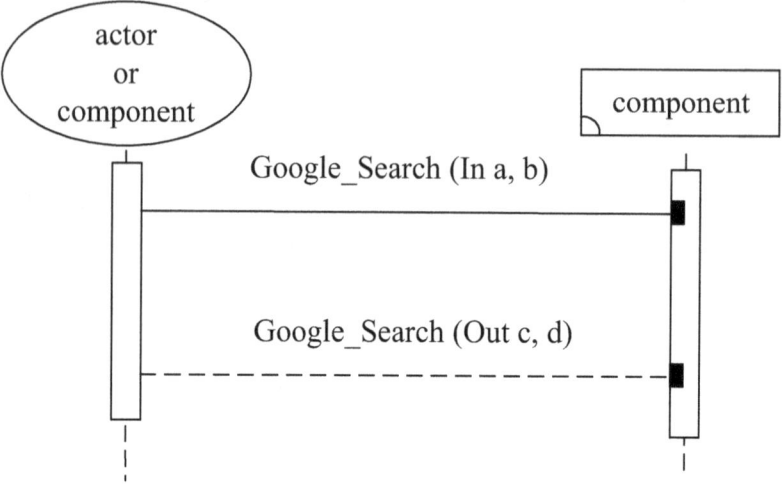

Or we can draw the operation-based interaction diagram as follows.

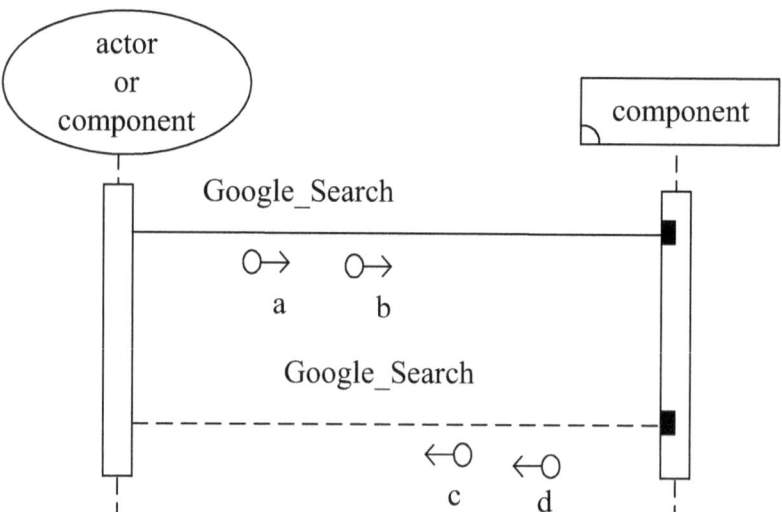

We use an operation-based internal interaction (i.e. λ) to represent their handshake or communication, if the caller agent and the callee agent are the same one.

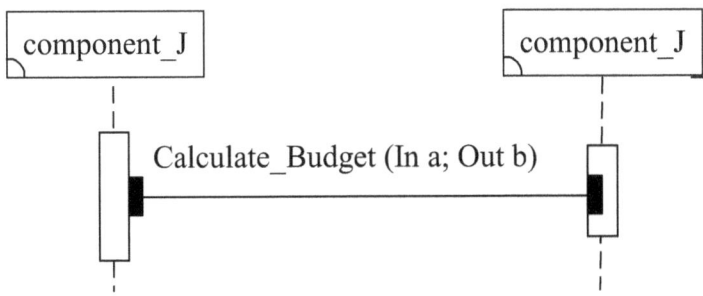

Also, we may redraw the operation-based internal interaction as follows.

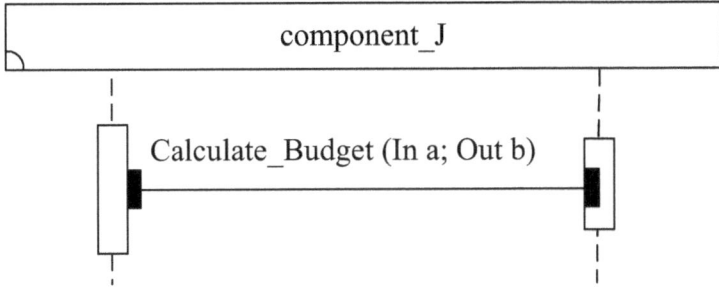

Formal Description of an Operation-Based Communication Port

We formally describe an operation-based communication port as a 3-tuple PORT = <operation_call_or_return, calling_or_called, operation_call_or_return_formula>, where "operation_call_or_return" stands for an OPERATION_CALL or OPERATION_RETURN tag, "calling_or_called" stands for a CALLING or CALLED action tag, and "operation_call_or_return_formula" stands for an operation call or operation return formula.

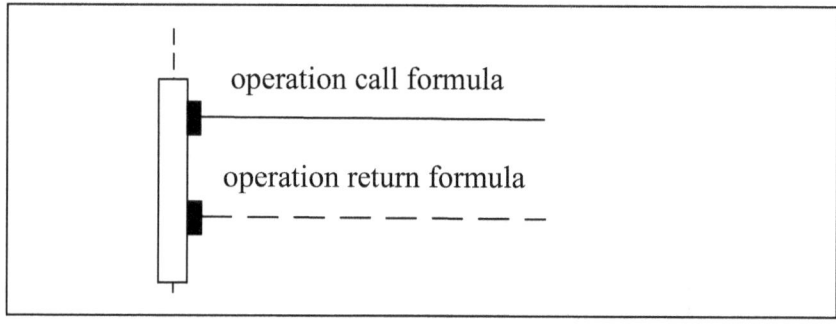

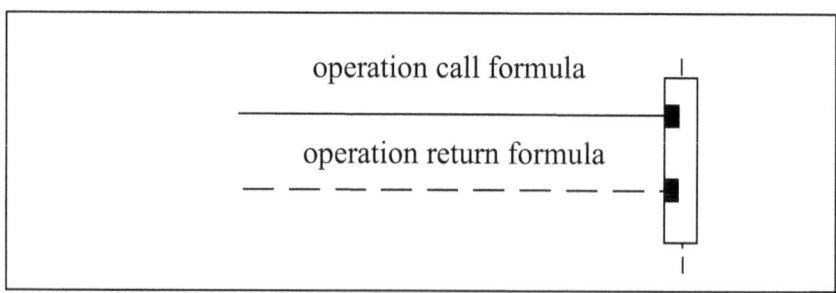

Formal Description of an Operation-Based Action

We formally describe an operation-based action as a 4-tuple ACTION = <operation_call_or_return, agent, calling_or_called, operation_call_or_return_formula>, where "operation_call_or_return" stands for an OPERATION_CALL or OPERATION_RETURN tag, "agent" stands for the name of a caller or callee agent, "calling_or_called" stands for a CALLING or CALLED action tag, and "operation_call_or_return_formula" stands for an operation call or operation return formula.

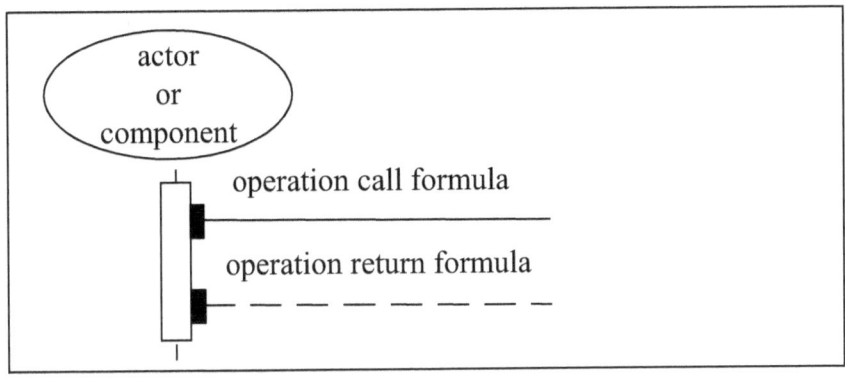

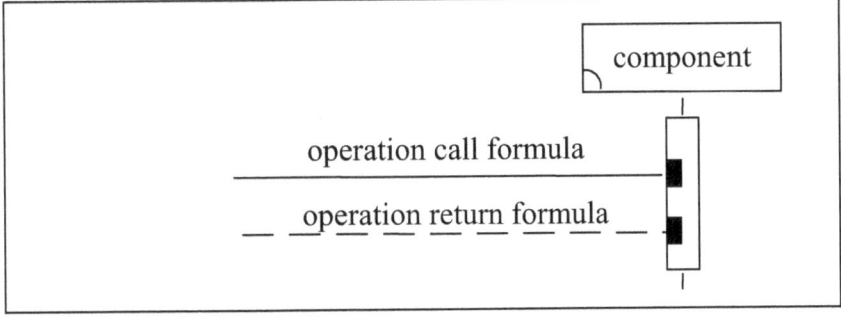

Formal Description of an Operation-Based Interaction

We formally describe an operation-based interaction as a 4-tuple INTERACTION = <operation_call_or_return, caller_agent, operation_call_or_return_formula, callee_agent>, where "operation_call_or_return" stands for an OPERATION_CALL or OPERATION_RETURN tag, "caller_agent" stands for the name of a caller agent, "operation_call_or_return_formula" stands for an operation call or operation return formula and "callee_agent" stands for the name of a callee agent.

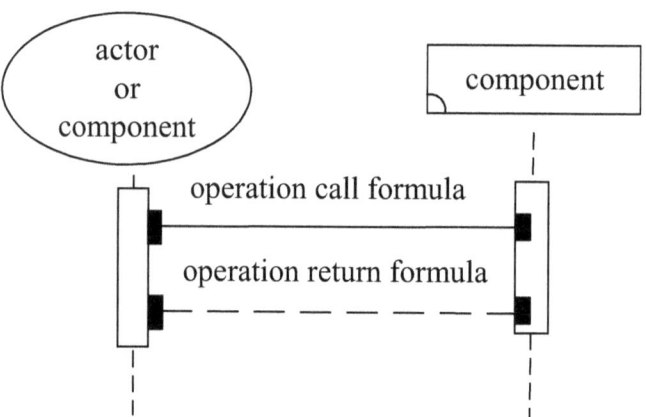

PART IV: MATHEMATICS OF GENERALIZED SBC PROCESSES

Sequentialization of Prefixes

Sometimes prefixes must be temporally ordered. For example, it might be desirable to specify algorithms such as: execute the "r" prefix (action/interaction with a condition) first and then execute the "P" process later. Sequentialization of prefixes can be used for such purposes.

Sequentialization of prefixes, usually written as the $r \bullet P$ process, indicates that it will perform the "r" prefix first and continue as the "P" process.

Summation of Processes

The binary operator "+", summation, combines two process expressions as alternatives.

For example, the process $P+Q$ can proceed non-deterministically either as the process P or the process Q; as soon as one performs its first action/interaction the other is discarded.

Parallel Composition of Processes

Parallel composition of two processes P and Q, usually written $P\|Q$, is the key primitive distinguishing the process algebras from sequential models of process executions.

Parallel composition allows the executions in P and Q to proceed simultaneously and independently.

Recursive Definition of a Process

The operators presented so far describe only finite action/interaction and are consequently insufficient for full computability, which includes non-terminating behavior. Recursion is the operator that allows finite descriptions of infinite behavior.

For example, **fix**$(X=z)$ can be understood as abbreviating the recursive definition of an infinite behavior denoted by the "X" process variable.

Replication of a Process

Replication is the other operator that allows finite descriptions of infinite behavior of a process.

For example, replication $!P$ can be understood as abbreviating the parallel composition of a countably infinite number of P processes.

Conditional Definition of a Process

A process can be defined by a one-or-more-armed conditional expression.

For example, the process (**if** $cond_1$ **then** m_1)+(**if** $cond_2$ **then** m_2)...+(**if** $cond_j$ **then** m_j) will proceed to perform the "m_1" action/interaction if the "$cond_1$" value is true, or proceed to perform the "m_2" action/interaction if the "$cond_2$" value is true,..., or proceed to perform the "m_j" action/interaction if the "$cond_j$" value is true.

.

Null Process

Process algebras generally also include a null process, denoted as *STOP* which has no interaction points. It is utterly inactive and its sole purpose is to act as the inductive anchor on top of which more interesting processes can be generated.

The process "*STOP•P₁*" (i.e. sequential composition of processes *STOP* and P_1) equals to the process "*STOP*".

$$STOP \bullet P_1 \quad = \quad STOP$$

The process "P_2+STOP" (i.e. summation of processes P_2 and *STOP*) equals to the process "*STOP+P₂*" (i.e. summation of processes *STOP* and P_2) which equals to the process "P_2".

$$P_2 + STOP \quad = \quad STOP + P_2 \quad = \quad P_2$$

The process "$P_3\|STOP$" (i.e. parallel composition of processes P_3 and $STOP$) equals to the process "$STOP\|P_3$" (i.e. parallel composition of processes $STOP$ and P_3) which equals to the process "P_3".

$$P_3 \parallel STOP \quad = \quad STOP \parallel P_3 \quad = \quad P_3$$

PART V: THE STRUCTURE-BEHAVIOR COALESCENCE APPROACH

Structure-Behavior Coalescence Means to Integrate the Systems Structure and Systems Behavior

Systems structure, specified by components, their channels and operations and their composition, refers to the type of connection between the components of a system.

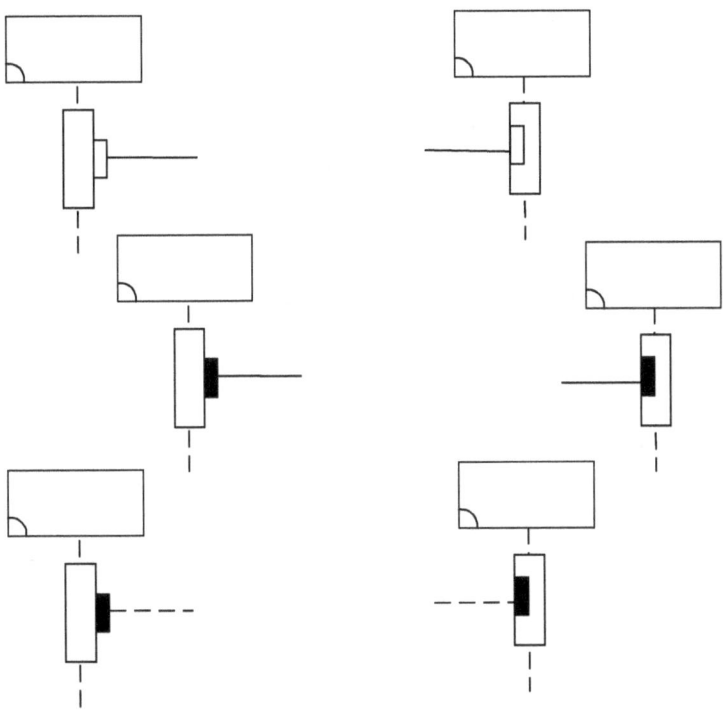

Systems behavior, specified by the interactions between and among the components and environment, refers to the interconnectivities a system in conjunction with its environment.

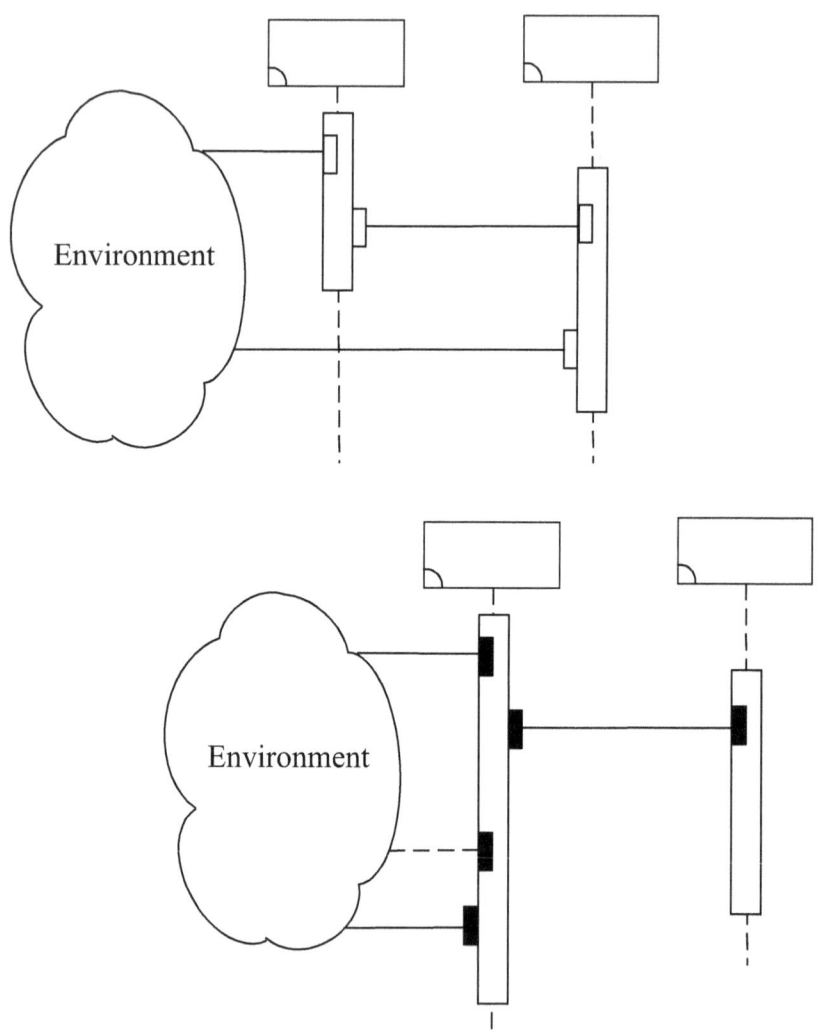

Systems structure and systems behavior are the two most prominent views of a system, integrating the systems structure and systems behavior is apparently the best way to achieve an integrated whole of a system.

If we are not able to integrate the systems structure and systems behavior, then there is no way that we are able to integrate the whole system.

Structure-behavior coalescence (SBC) [Chao15a] provides an elegant way to integrate the systems structure and systems behavior of a system. In other words, SBC facilitates an integrated whole of a system.

Interactions among Components and the External Environment to Draw Forth the Systems Behavior

All things that strike us as something independent are essentially parts of a system. We usually call the parts of a system its components. Components are sometimes labeled as parts, entities, objects, building blocks and non-aggregated systems [Chao14a, Chao14b, Chao14c, Chao16a].

In a system, if the components, and among them and the external environment to interact (or handshake), such interaction will draw forth the systems behavior.

A component uses an "action" to interact with the external environment as shown below.

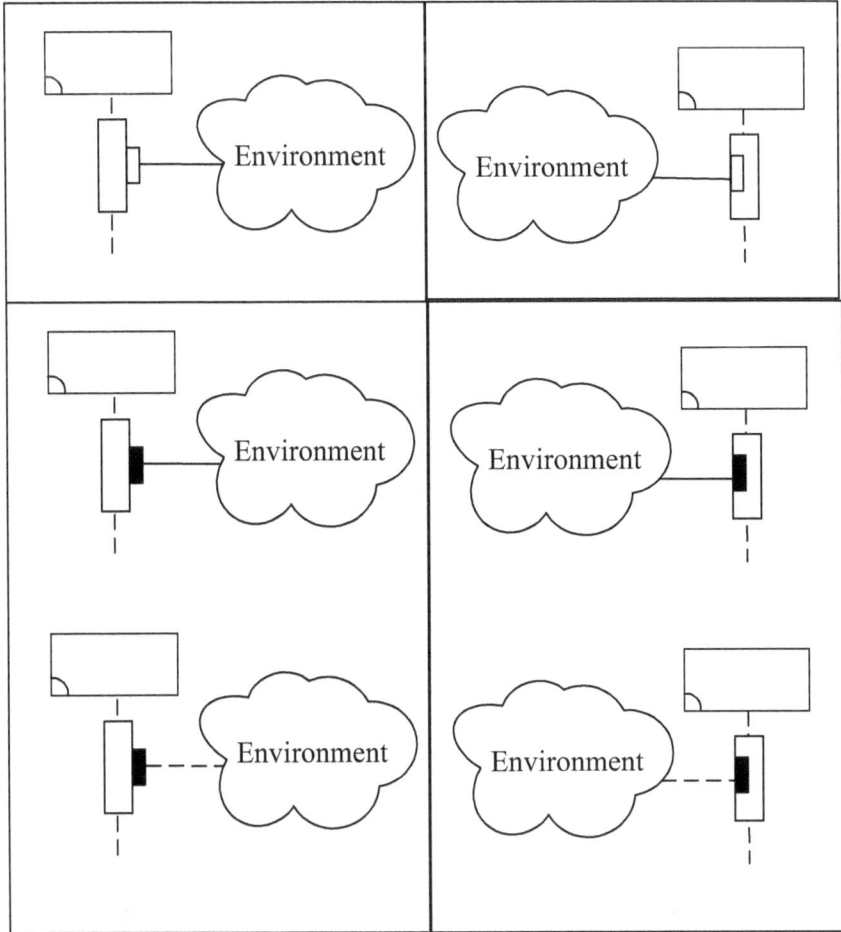

Two components use an "interaction" to interact with each other as shown below.

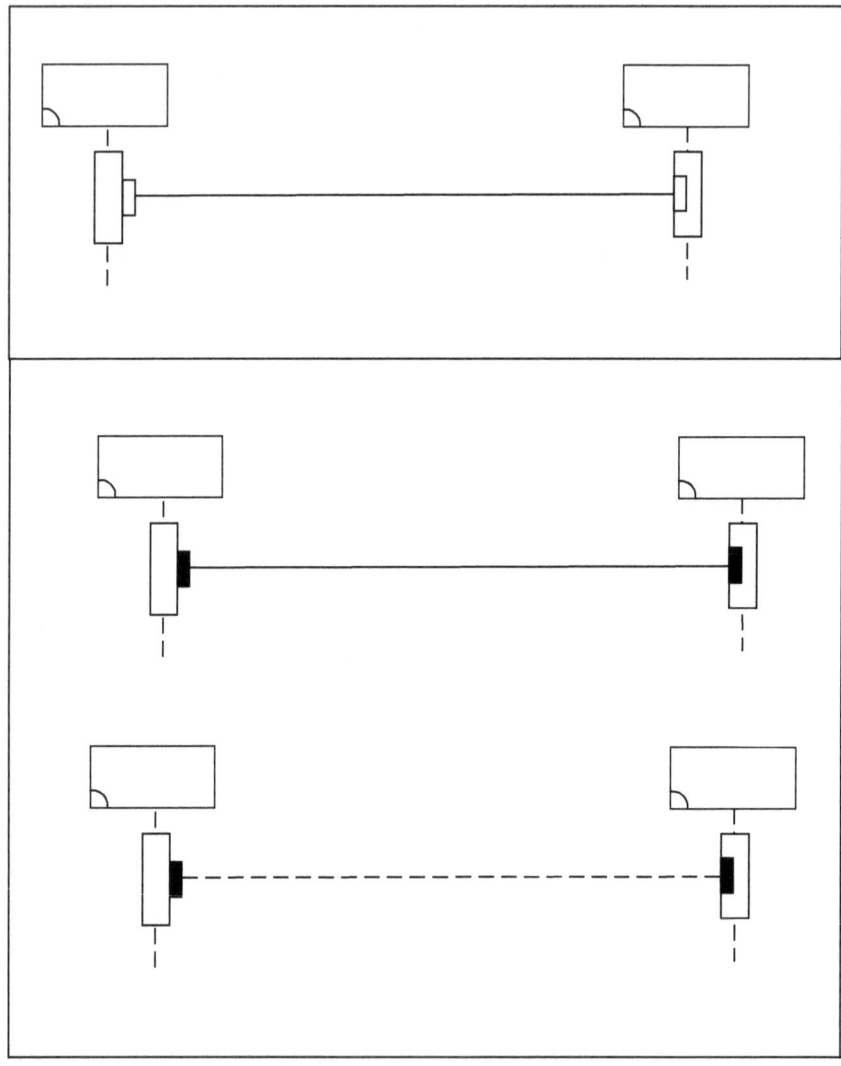

Core Theme of Structure-Behavior Coalescence

The core theme of structure-behavior coalescence is: "Systems Architecture = Systems Structure + Systems Behavior." That is, the systems structure will lead to the systems behavior.

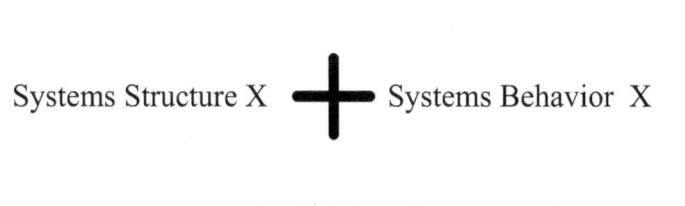

One systems structure will draw forth one systems behavior. That is, the systems behavior is attached to or built on the systems structure in the SBC approach.

In other words, the systems behavior can not exist alone; it must be loaded on the systems structure just like a cargo is loaded on a ship.

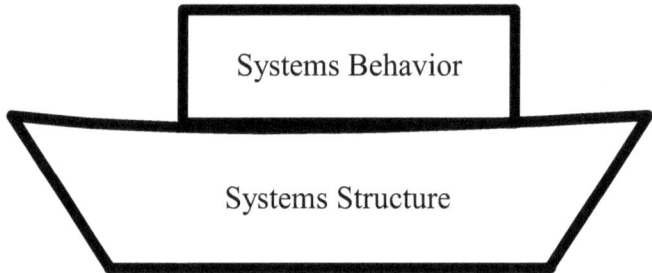

PART VI: LANGUAGE
CONSTRUCTS OF GENERALIZED
SBC PROCESS ALGEBRA

Entity Set and Entity Name

We assume an infinite set K of channel formulas, and use k to range over K. Further, we assume an infinite set L of operation call or operation return formulas, and use l to range over L. We let H be the set of actions, and use h to range over H. We let V be the set of non-internal interactions, and use v to range over V. Henceforward we let M be the set of all actions/interactions, and use m to range over M. We let R be the set of all condition actions/interactions, and use r to range over R. Further, we let X be the set of process variables, and use X_1, X_2 to range over X. We let Φ be the set of process Constants, and use A_1, A_2 to range over Φ. We let Π be the set of processes, and use P_1, Q_1 to range over Π. We let Ψ be the set of process expressions, and use E_1, E_2 to range over Ψ. Finally, we let Γ be the set of components, and use C_1, C_2 to range over Γ.

Entity set	Entity name	Type of entity
K	$k,...$	channel formulas
L	$l,...$	operation call or operation return formulas
H	$h,...$	actions
V	$v,...$	non-internal interactions
	λ	internal interaction
M	$m,...$	actions or non-internal interactions or internal interaction (actions/interactions)
R	$r,...$	condition actions/interactions (action/interaction with a condition)
R^*	$s,...$	condition action/interaction sequences
	$f,...$	renaming functions
X	$X_1, X_2,...$	process variables
Φ	$A_1, A_2,...$	process Constants
	$I, J,...$	indexing sets
Π	$P_1, Q_1,...$	processes
Ψ	$E_1, E_2,...$	process expressions
Γ	$C_1, C_2,...$	components
	S	bisimulations

Examples of Entities

As a first example, the r_{001} prefix defined as "$<C_{001}$, CG, $k_{001}>$" or "**if** *true* **then** $<C_{001}$, CG, $k_{001}>$" is a channel-based calling action under a tautology condition in which "C_{001}" stands for a caller component, "CG" stands for a calling action tag, and "k_{001}" stands for a channel formula.

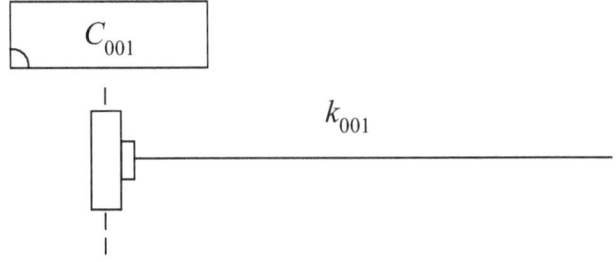

As a second example, the r_{002} prefix defined as "**if** $cond_{002}$ **then** $<C_{002}$, CG, $k_{002}>$" is a channel-based calling action under a certain (tautology excluded) condition in which "$cond_{002}$" stands for a (tautology excluded) condition, "C_{002}" stands for a caller component, "CG" stands for a calling action tag, and "k_{002}" stands for a channel formula.

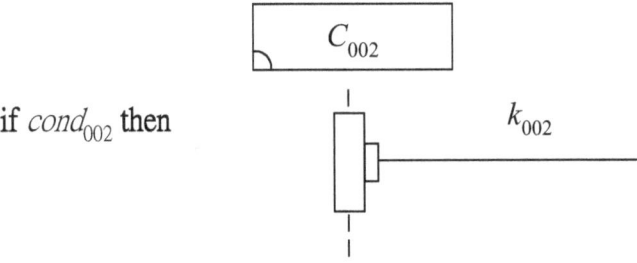

As a third example, the r_{003} prefix defined as "<O_C, C_{003}, CG, l_{003}>" or "**if** *true* **then** <O_C, C_{003}, CG, l_{003}>" is an operation-based calling action under a tautology condition in which "O_C" stands for an operation call action tag, "C_{003}" stands for a caller component, "CG" stands for a calling action tag, and "l_{003}" stands for an operation call formula.

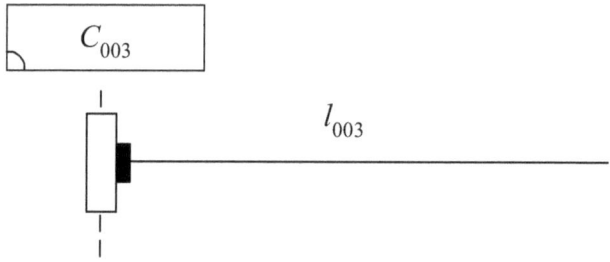

As a fourth example, the r_{004} prefix defined as "**if** *cond*$_{004}$ **then** <O_C, C_{004}, CG, l_{004}>" is an operation-based calling action under a certain (tautology excluded) condition in which "*cond*$_{004}$" stands for a (tautology excluded) condition, "O_C" stands for an operation call action tag, "C_{004}" stands for a caller component, "CG" stands for a calling action tag, and "l_{004}" stands for an operation call formula.

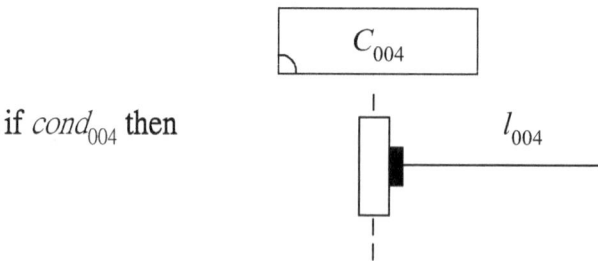

As a fifth example, the r_{005} prefix defined as "<O_R, C_{005}, CG, l_{005}>" or "**if** *true* **then** <O_R, C_{005}, CG, l_{005}>" is an operation-based calling action under a tautology condition in which "O_R" stands for an operation return action tag, "C_{005}" stands for a caller component, "CG" stands for a calling action tag, and "l_{005}" stands for an operation return formula.

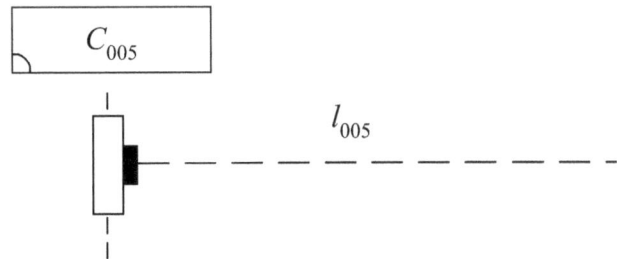

As a sixth example, the r_{006} prefix defined as "**if** *cond*$_{006}$ **then** <O_R, C_{006}, CG, l_{006}>" is an operation-based calling action under a certain (tautology excluded) condition in which "*cond*$_{006}$" stands for a (tautology excluded) condition, "O_R" stands for an operation return action tag, "C_{006}" stands for a caller component, "CG" stands for a calling action tag, and "l_{006}" stands for an operation return formula.

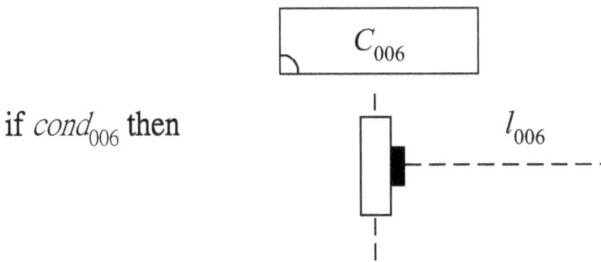

As a seventh example, the r_{007} prefix defined as "$<C_{007}$, CD, $k_{007}>$" or "**if** *true* **then** $<C_{007}$, CD, $k_{007}>$" is a channel-based called action under a tautology condition in which "C_{007}"stands for a callee component, "CD" stands for a called action tag, and "k_{007}" stands for a channel formula.

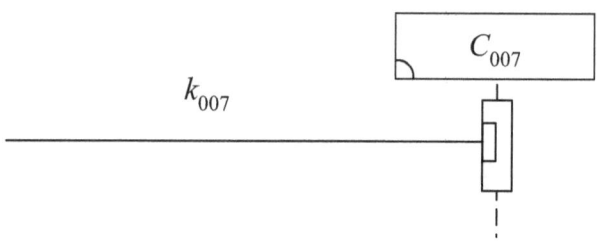

As a eighth example, the r_{008} prefix defined as **if** "$cond_{008}$ **then** $<C_{008}$, CD, $k_{008}>$" is a channel-based called action under a certain (tautology excluded) condition in which "$cond_{008}$" stands for a (tautology excluded) condition, "C_{008}" stands for a callee component, "CD" stands for a called action tag, and "k_{008}" stands for a channel formula.

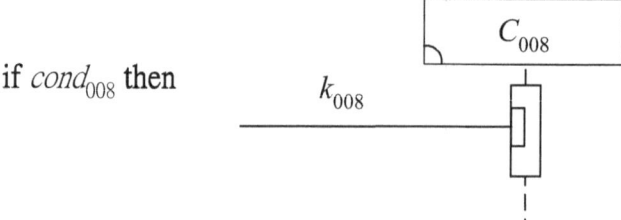

As a ninth example, the r_{009} prefix defined as "$<$O_C, C_{009}, CD, $l_{009}>$" or "**if** *true* **then** $<$O_C, C_{009}, CD, $l_{009}>$" is an operation-based called action under a tautology condition in which "O_C" stands for an operation call action tag, "C_{009}" stands for a callee component, "CD" stands for a called action tag, and "l_{009}" stands for an operation call formula.

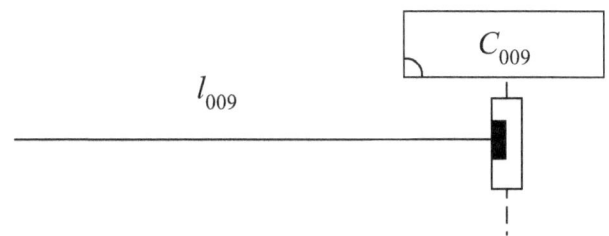

As a tenth example, the r_{010} prefix defined as "**if** $cond_{010}$ **then** $<$O_C, C_{010}, CD, $l_{010}>$" is an operation-based called action under a certain (tautology excluded) condition in which "$cond_{010}$" stands for a (tautology excluded) condition, "O_C" stands for an operation call action tag, "C_{010}" stands for a callee component, "CD" stands for a called action tag, and "l_{010}" stands for an operation call formula.

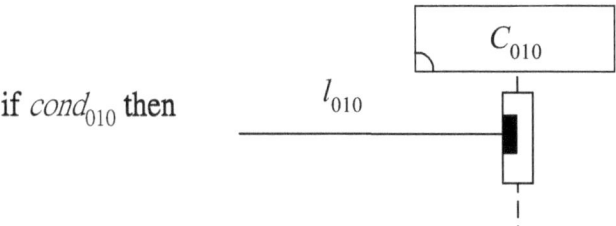

As a eleventh example, the r_{011} prefix defined as "<O_R C_{011}, CD, l_{011}>" or "**if** *true* **then** <O_R C_{011}, CD, l_{011}>" is an operation-based called action under a tautology condition in which "O_R" stands for an operation return action tag, "C_{011}" stands for a callee component, "CD" stands for a called action tag, and "l_{011}" stands for an operation return formula.

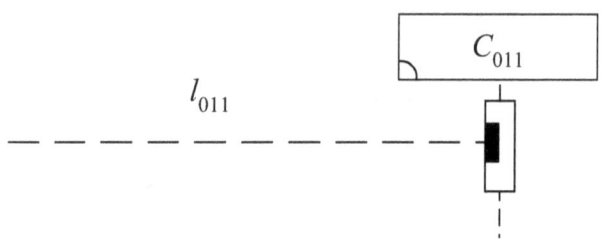

As a twelfth example, the r_{012} prefix defined as "**if** *cond*$_{012}$ **then** <O_R, C_{012}, CD, l_{012}>" is an operation-based called action under a certain (tautology excluded) condition in which "*cond*$_{012}$" stands for a (tautology excluded) condition, "O_R" stands for an operation return action tag, "C_{012}" stands for a callee component, "CD" stands for a called action tag, and "l_{012}" stands for an operation return formula.

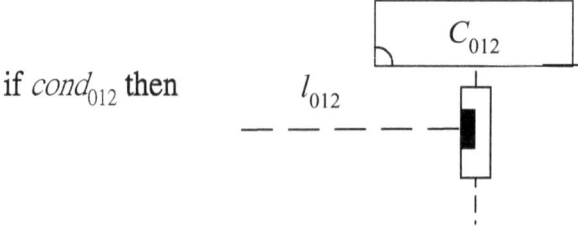

As a thirteenth example, the r_{013} prefix defined as "$<C_{013a}$, k_{013}, $C_{013b}>$" or "**if** *true* **then** $<C_{013a}, k_{013}, C_{013b}>$" is a channel-based interaction under a tautology condition in which "C_{013a}" stands for a caller component, "k_{013}" stands for a channel formula, and "C_{013b}" stands for a callee component.

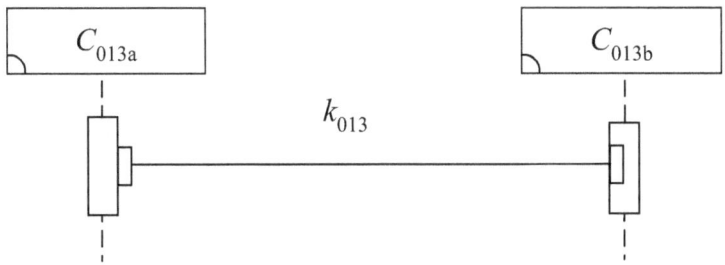

As a fourteenth example, the r_{014} prefix defined as "**if** *cond*$_{014}$ **then** $<C_{014a}, k_{014}, C_{014b}>$" is a channel-based interaction under a certain (tautology excluded) condition in which "*cond*$_{014}$" stands for a (tautology excluded) condition, "C_{014a}" stands for a caller component, "k_{014}" stands for a channel formula, and "C_{014b}" stands for a callee component.

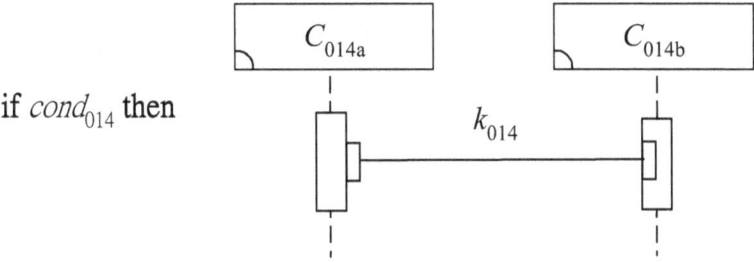

As an fifteenth example, the r_{015} prefix defined as "<O_C, C_{015a}, l_{015}, C_{015b}>" or "**if** *true* **then** <O_C, C_{015a}, l_{015}, C_{015b}>" is an operation-based interaction under a tautology condition in which "O_C" stands for an operation call interaction tag, "C_{015a}" stands for a caller component, "l_{015}" stands for an operation call formula, and "C_{015b}" stands for a callee component.

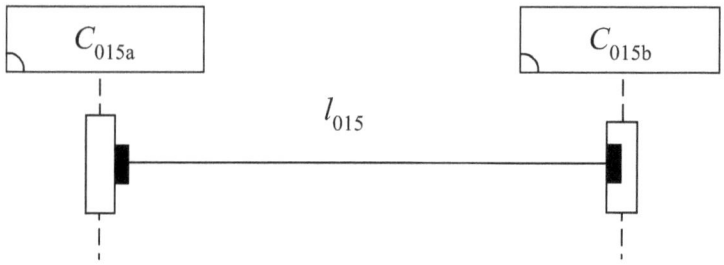

As an sixteenth example, the r_{016} prefix defined as "**if** *cond*$_{016}$ **then** <O_C, C_{016a}, l_{016}, C_{016b}>" is an operation-based interaction under a certain (tautology excluded) condition in which "*cond*$_{016}$" stands for a (tautology excluded) condition, "O_C" stands for an operation call interaction tag, "C_{016a}" stands for a caller component, "l_{016}" stands for an operation call formula, and "C_{016b}" stands for a callee component.

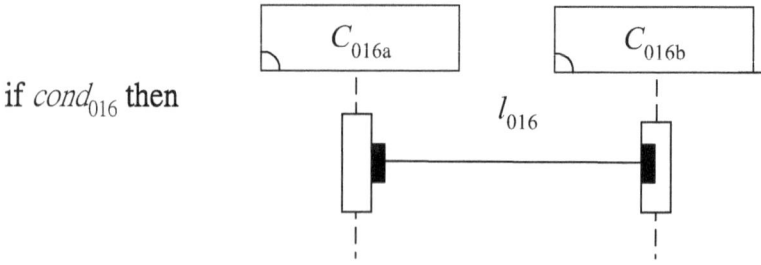

As a seventeenth example, the r_{017} prefix defined as "<O_R, C_{017a}, l_{017}, C_{017b}>" or "**if** *true* **then** <O_R, C_{017a}, l_{017}, C_{017b}>" is an operation-based interaction under a tautology condition in which "O_R" stands for an operation return interaction tag, "C_{017a}" stands for a caller component, "l_{017}" stands for an operation return formula, and "C_{017b}" stands for a callee component.

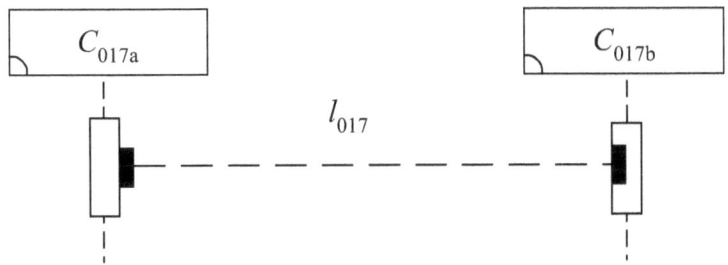

As a eighteenth example, the r_{018} prefix defined as "**if** $cond_{018}$ **then** <O_R, C_{018a}, l_{018}, C_{018b}>" is an operation-based interaction under a certain (tautology excluded) condition in which "$cond_{018}$" stands for a (tautology excluded) condition, "O_R" stands for an operation return interaction tag, "C_{018a}" stands for a caller component, "l_{018}" stands for an operation return formula, and "C_{018b}" stands for a callee component.

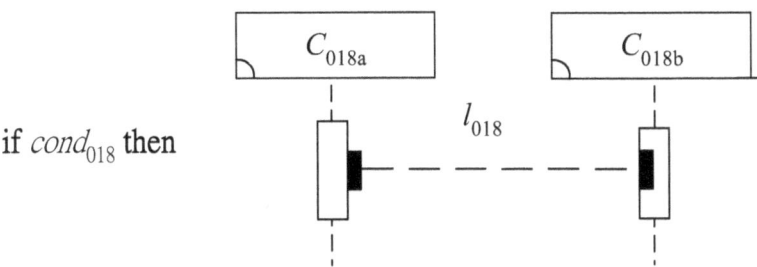

As a nineteenth example, the r_{019} prefix defined as "$<C_{019}$, k_{019}, $C_{019}>$" or "**if** *true* **then** $<C_{019}$, k_{019}, $C_{019}>$" is a channel-based internal interaction (i.e. λ) under a tautology condition in which "C_{019}" stands for a caller component, "k_{019}" stands for a channel formula, and "C_{019}" stands for a callee component. (Be noted that an internal interaction under a tautology condition is equal to an internal interaction.)

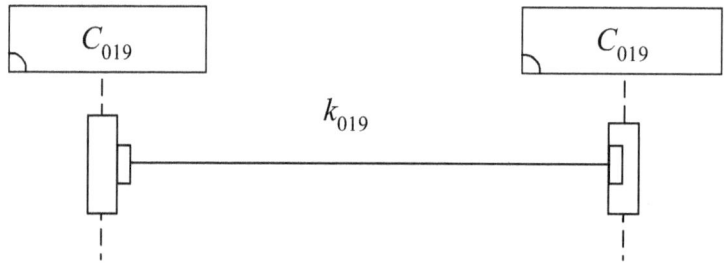

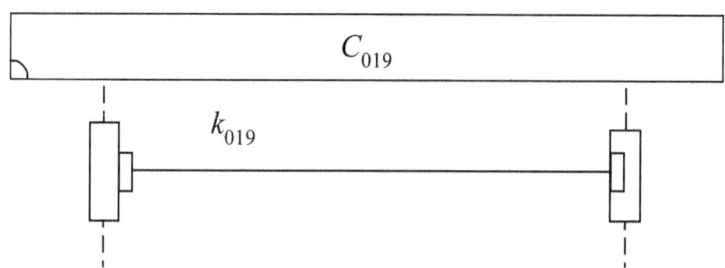

As a twentieth example, the r_{020} prefix defined as "**if** $cond_{020}$ **then** $<C_{020}, k_{020}, C_{020}>$" is a channel-based internal interaction (i.e. λ) under a certain (tautology excluded) condition in which "$cond_{020}$" stands for a (tautology excluded) condition, "C_{020}" stands for a caller component, "k_{020}" stands for a channel formula, and "C_{020}" stands for a callee component. (Be noted that an internal interaction under a certain (tautology excluded) condition is not equal to an internal interaction)

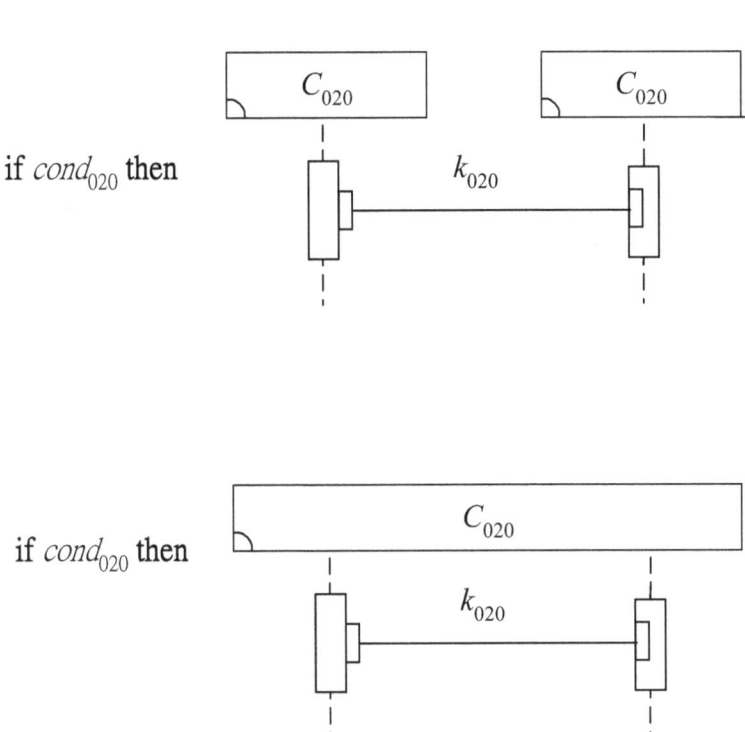

As an twenty-first example, the r_{021} prefix defined as "$<$O_C, C_{021}, l_{021}, $C_{021}>$" or "**if** *true* **then** $<$O_C, C_{021}, l_{021}, $C_{021}>$" is an operation-based internal interaction (i.e. λ) under a tautology condition in which "O_C" stands for an operation call interaction tag, "C_{021}" stands for a caller component, "l_{021}" stands for an operation call formula, and "C_{021}" stands for a callee component. (Be noted that an internal interaction under a tautology condition is equal to an internal interaction.)

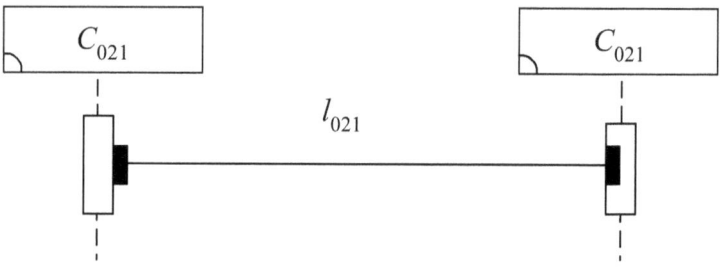

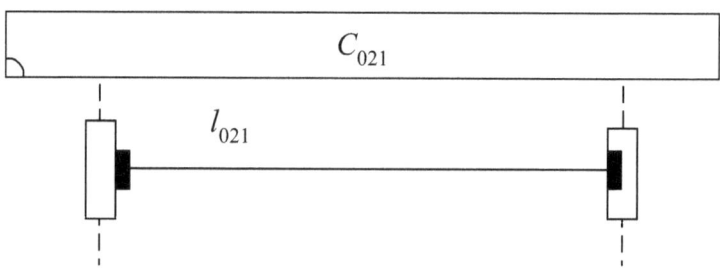

As an twenty-second example, the r_{022} prefix defined as "**if** $cond_{022}$ **then** <O_C, C_{022}, l_{022}, C_{022}>" is an operation-based internal interaction (i.e. λ) under a certain (tautology excluded) condition in which "$cond_{022}$" stands for a (tautology excluded) condition, "O_C" stands for an operation call interaction tag, "C_{022}" stands for a caller component, "l_{022}" stands for an operation call formula, and "C_{022}" stands for a callee component. (Be noted that an internal interaction under a certain (tautology excluded) condition is not equal to an internal interaction.)

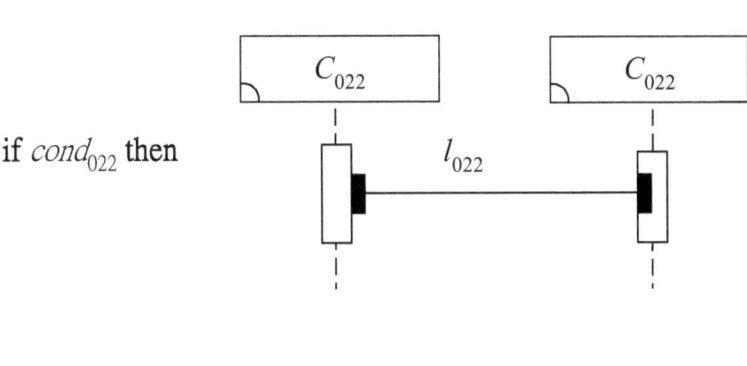

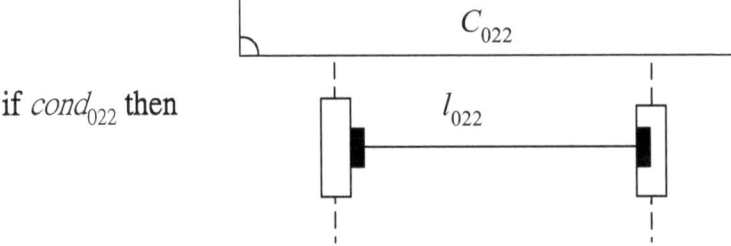

As a twenty-third example, the r_{023} prefix defined as "<O_R, C_{023}, l_{023}, C_{023}>" or "**if** *true* **then** <O_R, C_{023}, l_{023}, C_{023}>" is an operation-based internal interaction (i.e. λ) under a tautology condition in which "O_R" stands for an operation return interaction tag, "C_{023}" stands for a caller component, "l_{023}" stands for an operation return formula, and "C_{023}" stands for a callee component. (Be noted that an internal interaction under a tautology condition is equal to an internal interaction.)

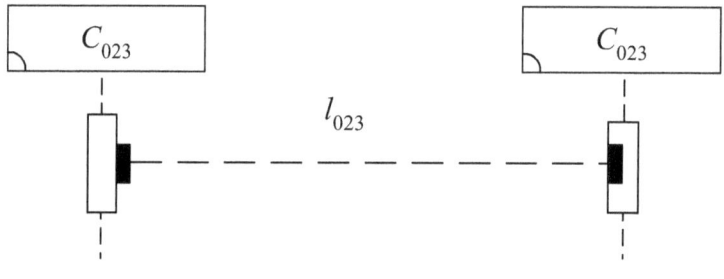

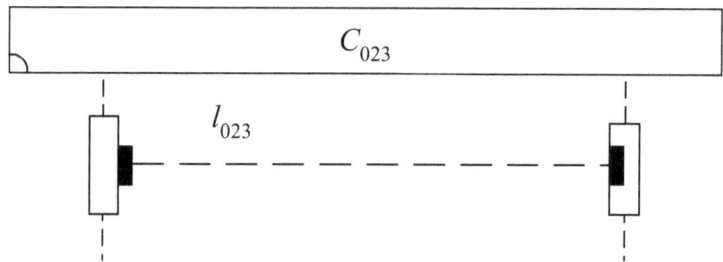

As a twenty-fourth example, the r_{024} prefix defined as "**if** $cond_{024}$ **then** <O_R, C_{024}, l_{024}, C_{024}>" is an operation-based internal interaction (i.e. λ) under a certain (tautology excluded) condition in which "$cond_{024}$" stands for a (tautology excluded) condition, "O_R" stands for an operation return interaction tag, "C_{024}" stands for a caller component, "l_{024}" stands for an operation return formula, and "C_{024}" stands for a callee component. (Be noted that an internal interaction under a certain (tautology excluded) condition is not equal to an internal interaction.)

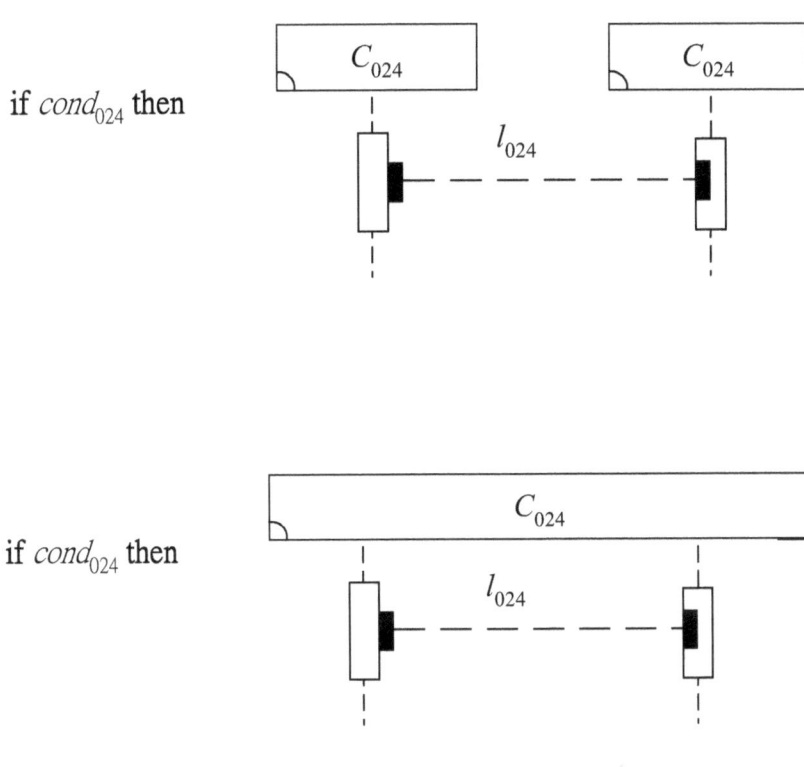

Backus-Naur Form of Generalized SBC Processes

The set of generalized SBC (i.e. Structure-Behavior Coalescence) processes is defined by the following BNF grammar:

(1) E ::= $STOP$

(2) | $r \bullet E$

(3) | $\sum_{i \in I} E_i$

(4) | $\mathbf{fix}(X=z)$

(5) | $E_1 \| E_2$

(6) | $E \setminus H$

(7) | $E[f]$

Null Expression

Rule 1 describes that the null process, denoted as *STOP* expression, is a valid generalized SBC process.

Rule 1
$E ::= STOP$

Prefix Expression

Rule 2 describes the process, denoted by the "$r \bullet E$" expression, performing the "r" (condition action/interaction) prefix and continuing as the process represented by the "E" expression, is a valid generalized SBC process.

Rule 2
$E \ ::= \ r \bullet E$

Summation Expression

Rule 3 describes the process, denoted by the $\sum_{i \in I} E_i$ expression, a sum of all expressions "E_i" as "i" ranges over "I" and can also be written as $\sum \{E_i : i \in I\}$, is a valid generalized SBC process.

Rule 3
$E ::= \sum_{i \in I} E_i$

Recursion Expression

Rule 4 describes the process, denoted by the "**fix**$(X=z)$" expression, a recursive definition of an infinite behavior represented by the process variable X, is a valid generalized SBC process.

Rule 4
$E \ ::= \ \textbf{fix} \ (X = z)$

Expression of Parallel Composition

Rule 5 describes the process, denoted by the $"E_1\|E_2"$ expression, parallel composition of two processes represented by the expressions $"E_1"$ and $"E_2"$, is a valid generalized SBC process.

Rule 5
$E \ ::= \ E_1 \| E_2$

Restriction Expression

Rule 6 describes the process, denoted by the expression ”$E\backslash H$” , hiding all actions (that appear in the set H) from the process represented by the expression ”E”, is a valid generalized SBC process.

Rule 6
$E \ ::= \ E \backslash H$

Renaming Expression

Rule 7 describes the process, denoted by the expression $E[f]$, having the effect of renaming the components (of the process represented by the expression "E") as dictated by "f", is a valid generalized SBC process.

Rule 7
$E \ ::= \ E[f]$

We often write $C'_1/C_1,\ldots, C'_n/C_n$ for the renaming function for which $f(C_i) = C'_i$ for $i = 1,\ldots, n$.

Examples of Generalized SBC Processes

As a first example, consider the generalized SBC process Constant A_{101} is defined as "$r_{101} \bullet r_{102} \bullet r_{103} \bullet STOP$". The Backus-Naur Form tree of A_{101} looks as follows:

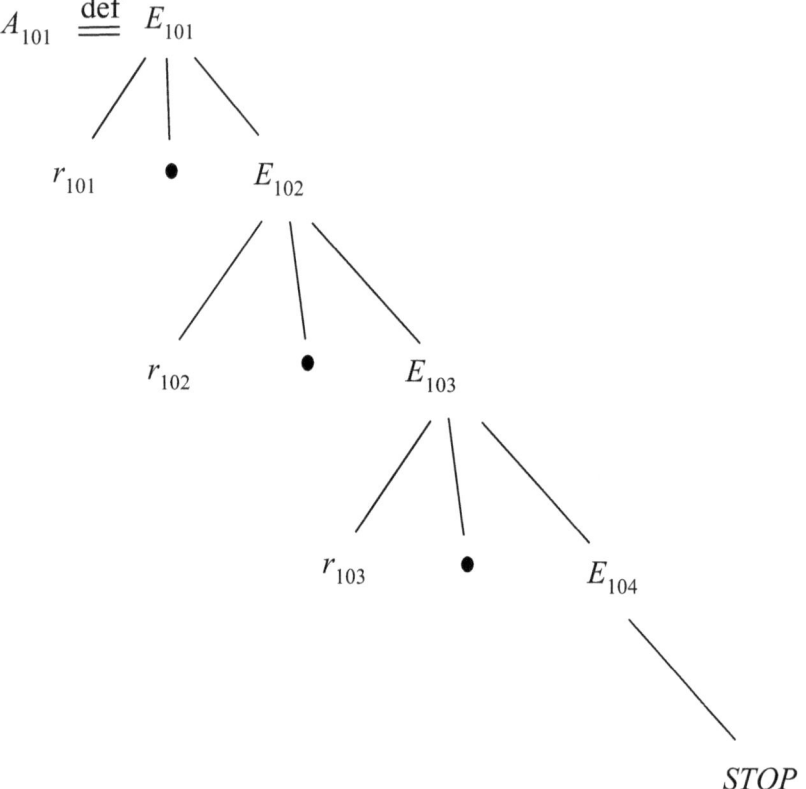

As a second example, consider the generalized SBC process Constant A_{102} is defined as "$r_{104} \bullet STOP + r_{105} \bullet STOP$" and the r_{104}, r_{105} prefixes are defined respectively as "$<C_{104}, CG, k_{104}>$", "$<C_{105}, CD, k_{105}>$". The Backus-Naur Form tree of A_{102} looks as follows:

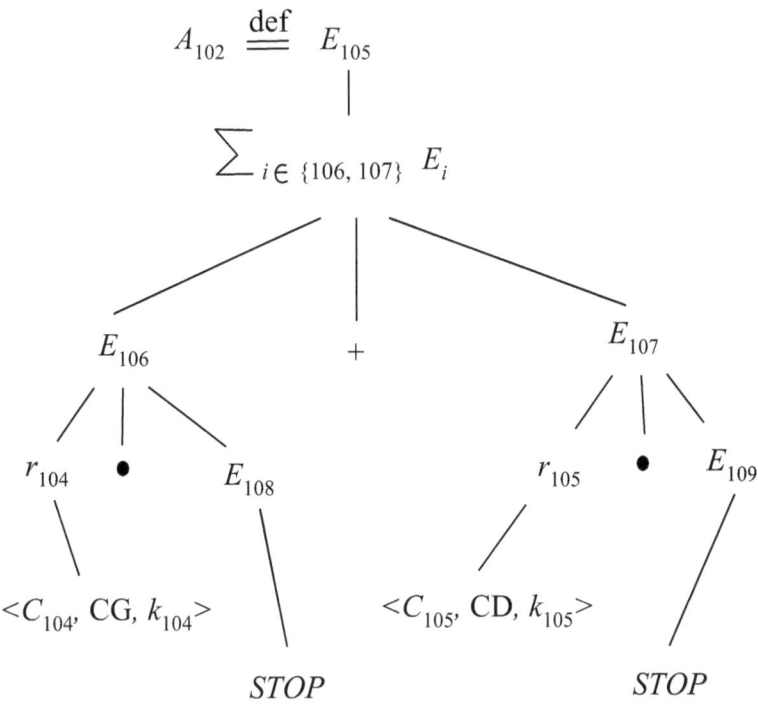

As a third example, consider the generalized SBC process Constant A_{103} is defined as "$r_{106} \bullet STOP + r_{107} \bullet STOP$" and the r_{106}, r_{107} prefixes are defined respectively as "$<C_{106},\ CG,\ k_{106}>$", "**if** $cond_{107}$ **then** $<C_{107},\ CD,\ k_{107}>$". The Backus-Naur Form tree of A_{103} looks as follows:

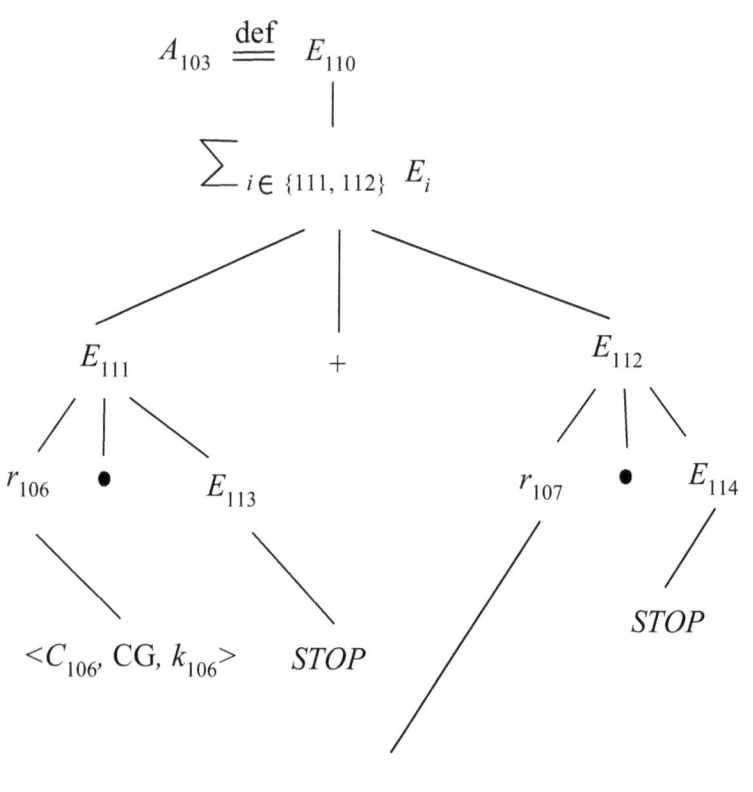

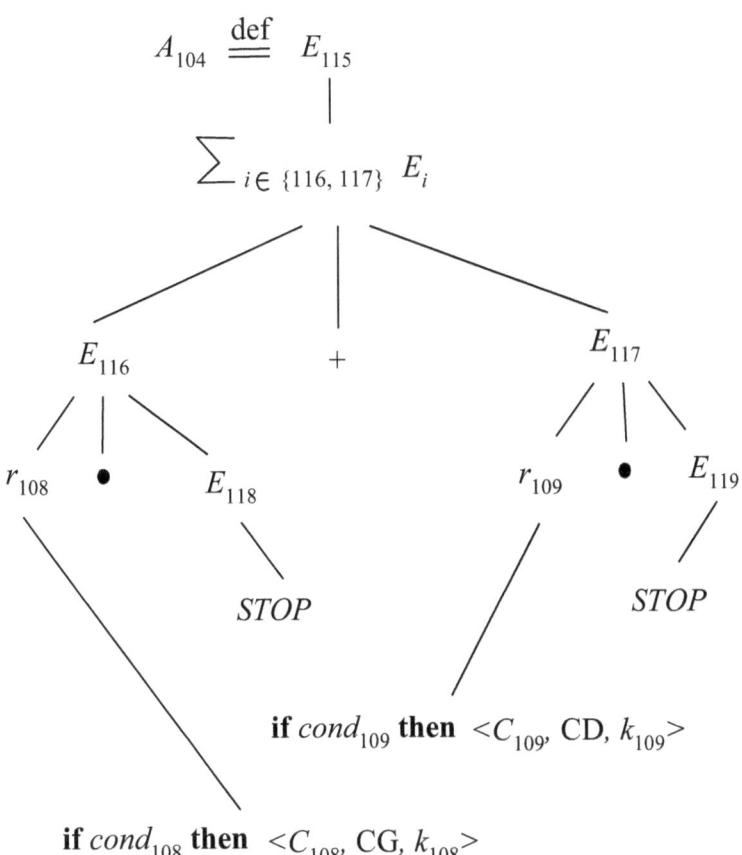

As a fourth example, consider the generalized SBC process Constant A_{104} is defined as "$r_{108} \bullet STOP + r_{109} \bullet STOP$" and the r_{108}, r_{109} prefixes are defined respectively as "**if** $cond_{108}$ **then** $<C_{108}$, CG, $k_{108}>$", "**if** $cond_{109}$ **then** $<C_{109}$, CD, $k_{109}>$". The Backus-Naur Form tree of A_{104} looks as follows:

As a fifth example, consider the generalized SBC process Constant A_{105} is defined as "**fix**$(X_{101}=r_{121}\bullet r_{122}\bullet X_{101}+r_{123}\bullet r_{124}\bullet X_{101})$". The Backus-Naur Form tree of A_{105} looks as follows:

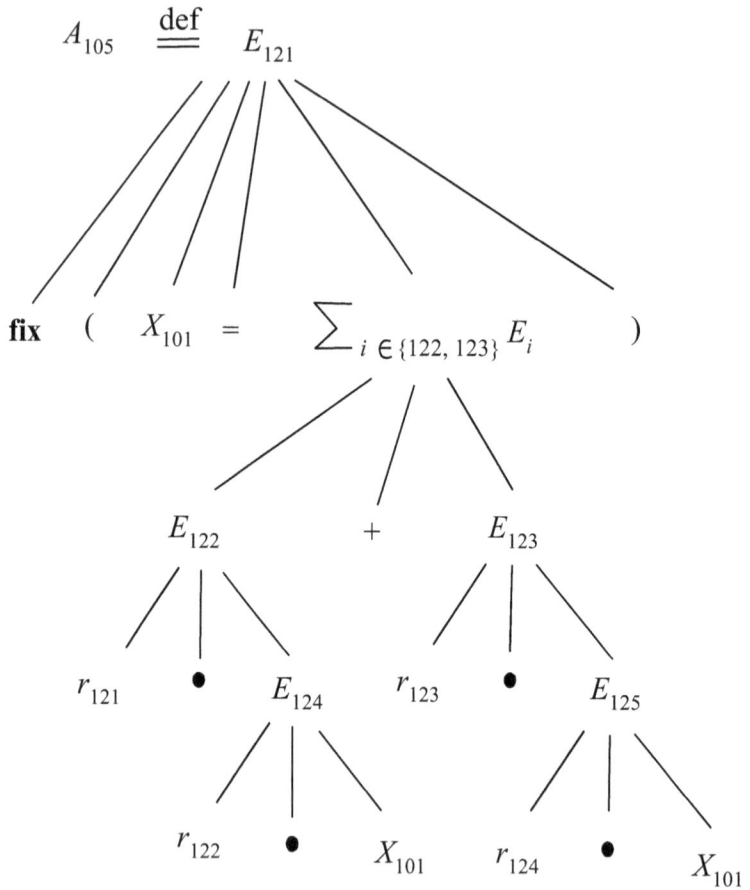

As a sixth example, consider the generalized SBC process Constant A_{106} is defined as "$r_{131} \bullet STOP \| r_{132} \bullet STOP$". The Backus-Naur Form tree of A_{106} looks as follows:

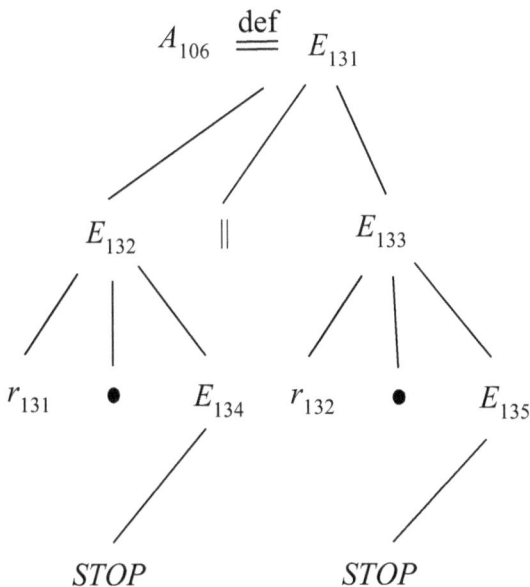

As a seventh example, consider the generalized SBC process Constant A_{107} is defined as "$r_{141} \bullet r_{142} \bullet r_{143} \bullet STOP \backslash \{r_{142}, r_{143}\}$". The Backus-Naur Form tree of A_{107} looks as follows:

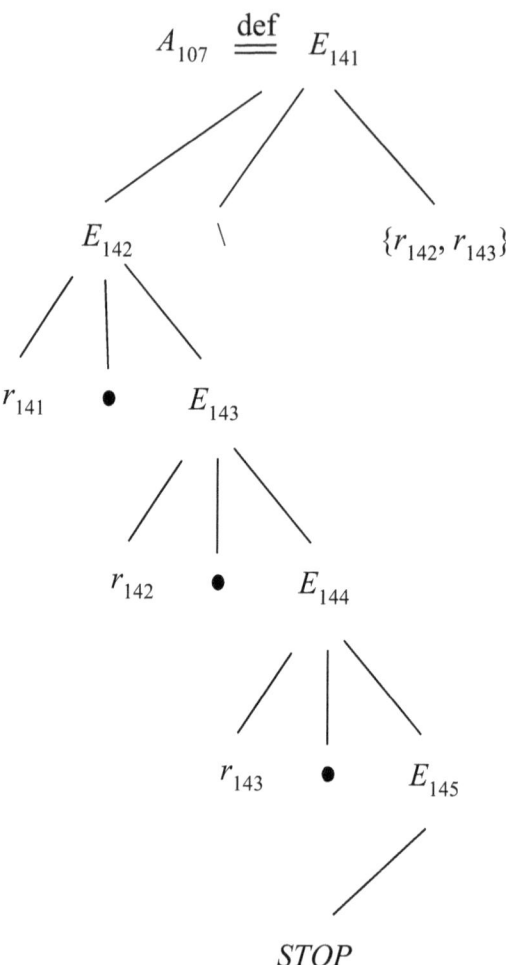

As a eighth example, consider the generalized SBC process Constant A_{108} is defined as "$r_{151} \bullet r_{152} \bullet STOP[C_{119}/C_{116}, C_{119}/C_{117}, C_{119}/C_{118}]$". The Backus-Naur Form tree of A_{108} looks as follows:

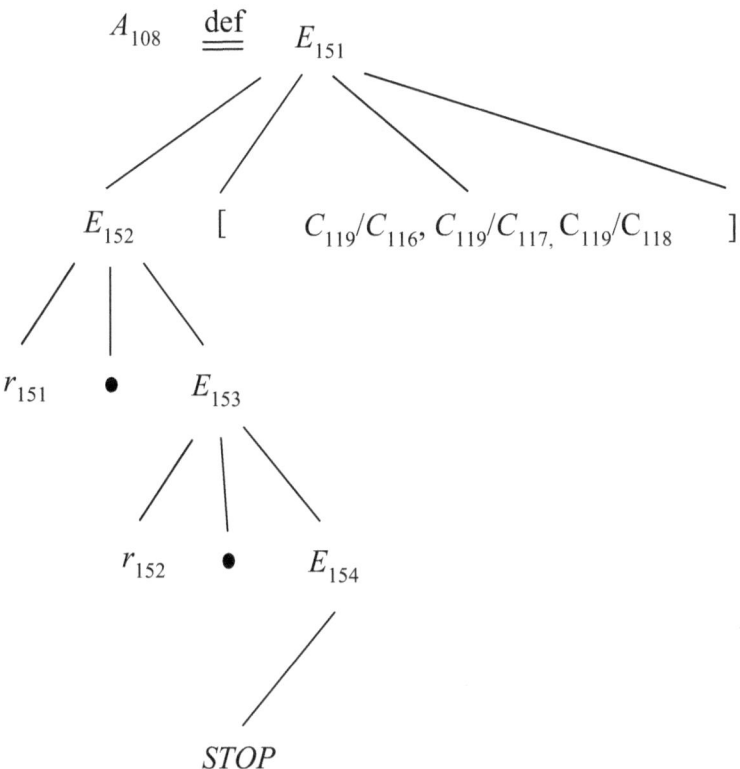

PART VII: TRANSITIONAL SEMANTICS OF GENERALIZED SBC PROCESS ALGEBRA

Transitional Semantics

In giving meaning to the generalized SBC process algebra, we shall use the following labelled transition system (LTS)

$$(\Psi, R, \rightarrow)$$

which consists of a set Ψ of process expressions, a set R of transition conditions/actions/interactions, and a transition relation

$\rightarrow \subseteq \Psi X R X \Psi$ where $(E_j, r, E_k) \in \rightarrow$ is written as $E_j \xrightarrow{r} E_k$.

The semantics for Ψ consists in the transition rules of each transition relation $\rightarrow$ over $\Psi X R X \Psi$. These transition rules will follow the structure of expressions.

We give the complete set of transition rules; the names **Prefix**, **Sum**, **Recursion**, **Parallel**, **Restriction**, **Rename**, and **Constant** indicate that the rules are associated respectively with Prefix, Summation, Recursion, Parallel Composition, Restriction, and Rename and with Constants.

Prefix
$$r \bullet E \xrightarrow{r} E$$

Sum$_j$
$$\frac{E_j \xrightarrow{r} E'_j}{\sum_{i \in I} E_i \xrightarrow{r} E'_j} (j \in I)$$

Recursion
$$\frac{\mathbf{fix}(X=z\{\mathbf{fix}(X=z)/X\}) \xrightarrow{r} E'}{\mathbf{fix}(X=z) \xrightarrow{r} E'}$$

Parallel$_1$
$$\frac{E \xrightarrow{r} E'}{E \parallel F \xrightarrow{r} E' \parallel F}$$

Parallel$_2$
$$\frac{F \xrightarrow{r} F'}{E \parallel F \xrightarrow{r} E \parallel F'}$$

Parallel$_3$

$$\frac{E \xrightarrow{\mathit{cond}_{121} <C_{121},\ \mathrm{CG},\ k>} E' \quad \bigwedge \quad F \xrightarrow{\mathit{cond}_{122} <C_{122},\ \mathrm{CD},\ k>} F'}{E \parallel F \xrightarrow{\mathit{cond}_{121} \& \mathit{cond}_{122} <C_{121},\ k,\ C_{122}>} E' \parallel F'}$$

Parallel$_4$

$$E \xrightarrow{\;cond_{123} <C_{123},\, \text{CD},\, k>\;} E' \;\bigwedge\; F \xrightarrow{\;cond_{124} <C_{124},\, \text{CG},\, k>\;} F'$$

$$E \parallel F \xrightarrow{\;cond_{123}\&cond_{124} <C_{124},\, k,\, C_{123}>\;} E' \parallel F'$$

Parallel$_5$

$$E \xrightarrow{\;cond_{125} <\text{O\_C},\, C_{125},\, \text{CG},\, l>\;} E' \;\bigwedge\; F \xrightarrow{\;cond_{126} <\text{O\_C},\, C_{126},\, \text{CD},\, l>\;} F'$$

$$E \parallel F \xrightarrow{\;cond_{125}\&cond_{126} <\text{O\_C},\, C_{125},\, l,\, C_{126}>\;} E' \parallel F'$$

Parallel$_6$

$$E \xrightarrow{\;cond_{127} <\text{O\_C},\, C_{127},\, \text{CD},\, l>\;} E' \;\bigwedge\; F \xrightarrow{\;cond_{128} <\text{O\_C},\, C_{128},\, \text{CG},\, l>\;} F'$$

$$E \parallel F \xrightarrow{\;cond_{127}\&cond_{128} <\text{O\_C},\, C_{128},\, l,\, C_{127}>\;} E' \parallel F'$$

Parallel$_7$

$$E \xrightarrow{\ cond_{129}\ {}^{<\text{O\_R},\ C_{129},\ \text{CG},\ l>}\ } E' \bigwedge F \xrightarrow{\ cond_{130}\ {}^{<\text{O\_R},\ C_{130},\ \text{CD},\ l>}\ } F'$$

$$E \parallel F \xrightarrow{\ cond_{129}\&cond_{130}\ {}^{<\text{O\_R},\ C_{129},\ l,\ C_{130}>}\ } E' \parallel F'$$

Parallel$_8$

$$E \xrightarrow{\ cond_{131}\ {}^{<\text{O\_R},\ C_{131},\ \text{CD},\ l>}\ } E' \bigwedge F \xrightarrow{\ cond_{132}\ {}^{<\text{O\_R},\ C_{132},\ \text{CG},\ l>}\ } F'$$

$$E \parallel F \xrightarrow{\ cond_{131}\&cond_{132}\ {}^{<\text{O\_R},\ C_{132},\ l,\ C_{131}>}\ } E' \parallel F'$$

Restriction $\qquad \dfrac{E \to E'}{E\backslash H \xrightarrow{r} E'\backslash H} \quad (r \notin H)$

Rename $\qquad \dfrac{E \xrightarrow{r} E'}{E[f] \xrightarrow{r[f]} E'[f]}$

Constant $\qquad \dfrac{P \xrightarrow{r} P'}{A \xrightarrow{r} P'} \quad (A \overset{\text{def}}{=\!=} P)$

Rule of Prefix

The rule for Prefix can be read as follows: Under any circumstances, we always infer $r \bullet E \xrightarrow{r} E$. That is, an expression, with a condition action/interaction prefixed to it, will use this condition action/interaction to accomplish the transition.

$$\frac{\phantom{r \bullet E \xrightarrow{r} E}}{r \bullet E \xrightarrow{r} E}$$

Rule of Summation

The rule for Summation can be read as follows: if any one summand E_j of the sum $\sum_{i \in I} E_i$ has a condition action/interaction, then the whole sum also has that condition action/interaction.

$$\frac{E_j \xrightarrow{r} E'_j}{\sum_{i \in I} E_i \xrightarrow{r} E'_j} \quad (j \in I)$$

Rule of Recursion

The rule for Recursion can be read as follows: This says that any condition action/interaction which may be inferred for the **fix** expression 'unwound' once (by substituting itself for its bound variable) may be inferred for the **fix** expression itself.

$$\frac{\mathbf{fix}(X{=}z\{\mathbf{fix}(X{=}z)/X\}) \overset{r}{\rightarrow} E'}{\mathbf{fix}(X{=}z) \overset{r}{\rightarrow} E'}$$

Rules of Parallel Composition

There are eight transition rules for parallel composition. Rule Parallel$_1$ indicates that from $E \xrightarrow{r} E'$ we shall infer $E\|F \xrightarrow{r} E'\|F$.

$$\frac{E \xrightarrow{r} E'}{E \| F \xrightarrow{r} E' \| F}$$

Rule Parallel$_2$ indicates that from $F \xrightarrow{r} F'$ we shall infer $E\|F \xrightarrow{r} E\|F'$.

$$\frac{F \xrightarrow{r} F'}{E \| F \xrightarrow{r} E \| F'}$$

Rule Parallel$_3$ indicates that from E $\xrightarrow{cond_{121} \; <C_{121}, \; \text{CG}, \; k>} E'$ and F $\xrightarrow{cond_{122} \; <C_{122}, \; \text{CD}, \; k>} F'$ we shall infer $E \| F$ $\xrightarrow{cond_{121} \& cond_{122} \; <C_{121}, \; k, \; C_{122}>} E' \| F'$.

$$E \xrightarrow{cond_{121} \; <C_{121}, \; \text{CG}, \; k>} E' \quad \bigwedge \quad F \xrightarrow{cond_{122} \; <C_{122}, \; \text{CD}, \; k>} F'$$

$$\overline{}$$

$$E \| F \xrightarrow{cond_{121} \& cond_{122} \; <C_{121}, \; k, \; C_{122}>} E' \| F'$$

Rule Parallel$_4$ indicates that from E $\xrightarrow{cond_{123} \; <C_{123}, \; \text{CD}, \; k>} E'$ and F $\xrightarrow{cond_{124} \; <C_{124}, \; \text{CG}, \; k>} F'$ we shall infer $E \| F$ $\xrightarrow{cond_{123} \& cond_{124} \; <C_{124}, \; k, \; C_{123}>} E' \| F'$.

$$E \xrightarrow{cond_{123} \; <C_{123}, \; \text{CD}, \; k>} E' \quad \bigwedge \quad F \xrightarrow{cond_{124} \; <C_{124}, \; \text{CG}, \; k>} F'$$

$$\overline{}$$

$$E \| F \xrightarrow{cond_{123} \& cond_{124} \; <C_{124}, \; k, \; C_{123}>} E' \| F'$$

Rule Parallel$_5$ indicates that from E $\xrightarrow{\ cond_{125}\ <\text{O\_C},\ C_{125},\ \text{CG},\ l>\ } E'$ and F $\xrightarrow{\ cond_{126}\ <\text{O\_C},\ C_{126},\ \text{CD},\ l>\ } F'$ we shall infer $E\|F$ $\xrightarrow{\ cond_{125}\&cond_{126}\ <\text{O\_C},\ C_{125},\ l,\ C_{126}>\ } E'\|F'.$

$$\frac{E \xrightarrow{\ cond_{125}\ <\text{O\_C},\ C_{125},\ \text{CG},\ l>\ } E' \quad \bigwedge \quad F \xrightarrow{\ cond_{126}\ <\text{O\_C},\ C_{126},\ \text{CD},\ l>\ } F'}{E\|F \xrightarrow{\ cond_{125}\&cond_{126}\ <\text{O\_C},\ C_{125},\ l,\ C_{126}>\ } E'\|F'}$$

Rule Parallel$_6$ indicates that from E $\xrightarrow{\quad cond_{127} <\text{O\_C}, C_{127}, \text{CD}, l> \quad}$ E' and F $\xrightarrow{\quad cond_{128} <\text{O\_C}, C_{128}, \text{CG}, l> \quad}$ F' we shall infer $E\|F$ $\xrightarrow{\quad cond_{127}\&cond_{128} <\text{O\_C}, C_{128}, l, C_{127}> \quad}$ $E'\|F'$.

$$\frac{E \xrightarrow{\quad cond_{127} <\text{O\_C}, C_{127}, \text{CD}, l> \quad} E' \quad\bigwedge\quad F \xrightarrow{\quad cond_{128} <\text{O\_C}, C_{128}, \text{CG}, l> \quad} F'}{E \| F \xrightarrow{\quad cond_{127}\&cond_{128} <\text{O\_C}, C_{128}, l, C_{127}> \quad} E' \| F'}$$

Rule Parallel$_7$ indicates that from E $\xrightarrow{cond_{129}\ <\text{O\_R},\ C_{129},\ \text{CG},\ l>}$ E' and F $\xrightarrow{cond_{130}\ <\text{O\_R},\ C_{130},\ \text{CD},\ l>}$ F' we shall infer $E\|F$ $\xrightarrow{cond_{129}\&cond_{130}\ <\text{O\_R},\ C_{129},\ l,\ C_{130}>}$ E'$\|F$'.

$$\frac{E \xrightarrow{cond_{129}\ <\text{O\_R},\ C_{129},\ \text{CG},\ l>} E' \bigwedge F \xrightarrow{cond_{130}\ <\text{O\_R},\ C_{130},\ \text{CD},\ l>} F'}{E\,\|\,F \xrightarrow{\quad cond_{129}\&cond_{130}\ <\text{O\_R},\ C_{129},\ l,\ C_{130}>\quad} E'\,\|\,F'}$$

Rule Parallel$_8$ indicates that from E $\xrightarrow{cond_{131} <\text{O\_R}, C_{131}, \text{CD}, l>} E'$ and F $\xrightarrow{cond_{132} <\text{O\_R}, C_{132}, \text{CG}, l>} F'$ we shall infer $E\|F$ $\xrightarrow{cond_{131}\&cond_{132} <\text{O\_R}, C_{132}, l, C_{131}>} E'\|F'$.

$$\frac{E \xrightarrow{cond_{131} <\text{O\_R}, C_{131}, \text{CD}, l>} E' \bigwedge F \xrightarrow{cond_{132} <\text{O\_R}, C_{132}, \text{CG}, l>} F'}{E\|F \xrightarrow{cond_{131}\&cond_{132} <\text{O\_R}, C_{132}, l, C_{131}>} E'\|F'}$$

Rule of Restriction

The rule for Restriction can be read as follows: Rule Restriction indicates that from $E \xrightarrow{r} E'$ we shall infer $E \backslash H \xrightarrow{r} E' \backslash H$ as only as $r \notin H$.

$$\frac{E \xrightarrow{r} E'}{E \backslash H \xrightarrow{r} E' \backslash H} \quad (r \notin H)$$

Rule of Rename

The rule for Rename can be read as follows: Rule Rename indicates that from $E \xrightarrow{r} E'$ we shall infer $E[f] \xrightarrow{r[f]} E'[f]$.

$$
\frac{E \xrightarrow{r} E'}{E[f] \xrightarrow{r} E'[f]}
$$

Rule of Constants

The rule for Constants can be read as follows: the rule of Constants asserts that each Constant has the same condition actions/transitions as its defining expression.

$$\frac{P \xrightarrow{r} P'}{A \xrightarrow{r} P'} \quad (A \overset{\text{def}}{=} P)$$

Examples of Transitional Semantics

As a first example, consider the generalized SBC process Constant A_{111} is defined as "$r_{201} \bullet r_{202} \bullet r_{203} \bullet STOP$". The following transition graph shows the semantics of process A_{111}.

$$A_{111} \xrightarrow{\; r_{201} \;} P_{111} \xrightarrow{\; r_{202} \;} P_{112} \xrightarrow{\; r_{203} \;} STOP$$

In the transition graph of the A_{111}'s generalized SBC process, processes A_{111}, P_{111} and P_{112} are defined as:

$$A_{111} \stackrel{\text{def}}{=\!=} r_{201} \bullet P_{111}$$

$$P_{111} \stackrel{\text{def}}{=\!=} r_{202} \bullet P_{112}$$

$$P_{112} \stackrel{\text{def}}{=\!=} r_{203} \bullet STOP$$

As a second example, consider the generalized SBC process Constant A_{112} is defined as "$(r_{204} \bullet r_{205} \bullet STOP) + (r_{206} \bullet r_{207} \bullet STOP)$". The following transition graph shows the semantics of process A_{112}.

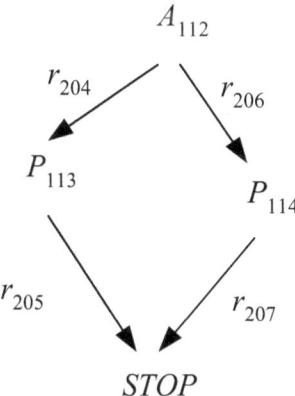

In the transition graph of the A_{112}'s generalized SBC process, processes A_{112}, P_{113} and P_{114} are defined as:

$$A_{112} \overset{\text{def}}{=\!=} r_{204} \bullet P_{113} + r_{206} \bullet P_{114}$$

$$P_{113} \overset{\text{def}}{=\!=} r_{205} \bullet STOP$$

$$P_{114} \overset{\text{def}}{=\!=} r_{207} \bullet STOP$$

As a third example, consider the generalized SBC process Constant A_{113} is defined as "(**if** $cond_{208}$ **then** $m_{208} \bullet m_{209} \bullet STOP$)+(**if** $cond_{210}$ **then** $m_{210} \bullet m_{211} \bullet STOP$)". The following transition graph shows the semantics of process A_{113}.

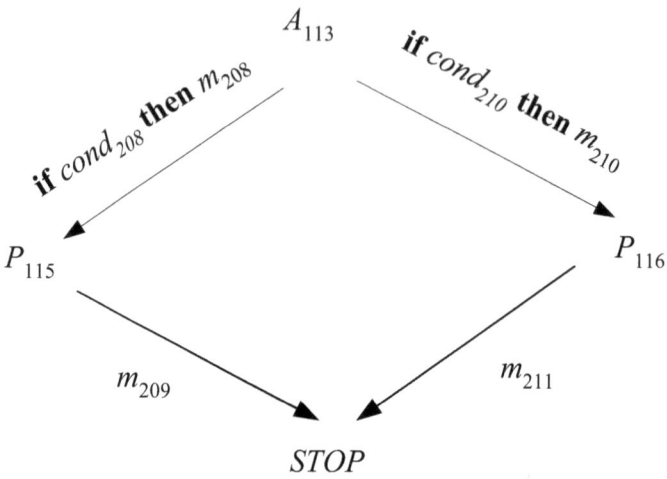

In the transition graph of the A_{113}'s generalized SBC process, processes A_{113}, P_{115} and P_{116} are defined as:

$$A_{113} \stackrel{\text{def}}{=\!=} \text{ if } cond_{208} \text{ then } m_{208} \bullet P_{115} + \text{if } cond_{210} \text{ then } m_{210} \bullet P_{116}$$

$$P_{115} \stackrel{\text{def}}{=\!=} m_{209} \bullet STOP$$

$$P_{116} \stackrel{\text{def}}{=\!=} m_{211} \bullet STOP$$

As a fourth example, consider the generalized SBC process Constant A_{114} is defined as "$\mathbf{fix}(X_{114}=r_{212}\bullet r_{213}\bullet r_{214}\bullet X_{114})$". The following transition graph shows the semantics of process A_{114}.

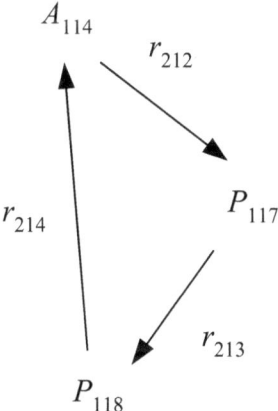

In the transition graph of the A_{114}'s generalized SBC process, processes A_{114}, P_{117} and P_{118} are defined as:

$$A_{114} \overset{\text{def}}{=\!=} r_{212} \bullet P_{117}$$

$$P_{117} \overset{\text{def}}{=\!=} r_{213} \bullet P_{118}$$

$$P_{118} \overset{\text{def}}{=\!=} r_{214} \bullet A_{114}$$

As a fifth example, consider the generalized SBC process Constant A_{115} is defined as "$\mathbf{fix}(X_{115}=r_{215}\bullet r_{216}\bullet X_{115}+r_{217}\bullet r_{218}\bullet X_{115})$". The following transition graph shows the semantics of process A_{115}.

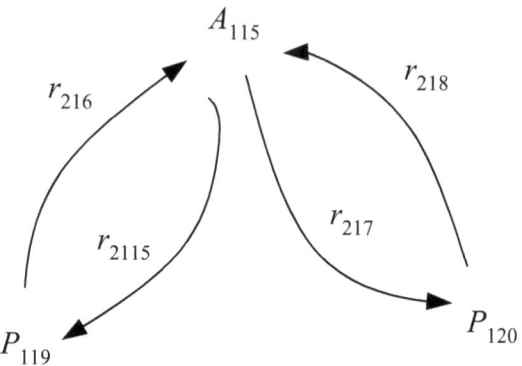

In the transition graph of the A_{115}'s generalized SBC process, processes A_{115}, P_{119} and P_{120} are defined as:

$$A_{115} \stackrel{\text{def}}{=\!=} r_{215} \bullet P_{119} + r_{217} \bullet P_{120}$$

$$P_{119} \stackrel{\text{def}}{=\!=} r_{216} \bullet A_{115}$$

$$P_{120} \stackrel{\text{def}}{=\!=} r_{218} \bullet A_{115}$$

As a sixth example, consider the generalized SBC process Constant A_{116} is defined as "**fix**$(X_{116}=r_{219} \bullet r_{220} \bullet X_{116}+r_{221} \bullet r_{222} \bullet r_{223} \bullet X_{116})$". The following transition graph shows the semantics of process A_{116}.

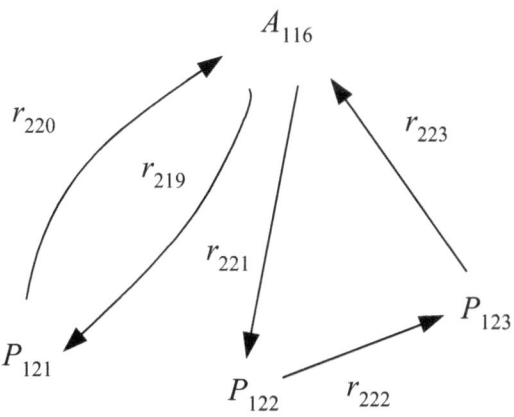

In the transition graph of the A_{116}'s generalized SBC process, processes A_{116}, P_{121}, P_{122} and P_{123} are defined as:

$$A_{116} \stackrel{\text{def}}{=\joinrel=} r_{219} \bullet P_{121} + r_{221} \bullet P_{122}$$

$$P_{121} \stackrel{\text{def}}{=\joinrel=} r_{220} \bullet A_{116}$$

$$P_{122} \stackrel{\text{def}}{=\joinrel=} r_{222} \bullet P_{123}$$

$$P_{123} \stackrel{\text{def}}{=\joinrel=} r_{223} \bullet A_{116}$$

PART VIII: EXAMPLES OF INTERLEAVING

Interleaving

Interleaving is a kind of parallel composition which represents completely independent concurrent activity of involved processes [Hoar85]. For example, if we interleave process P_1 with process P_2 then the activities from both processes will be arbitrarily interleaved in time.

In general, only transition rule Parallel$_1$ or Parallel$_2$ is applied when several processes are interleaved. That is, we shall not use transition rules Parallel$_3$, Parallel$_4$, Parallel$_5$, Paralle$_6$, Parallel$_7$ and Parallel$_8$ for the purpose of interleaving.

Channel-Based Interleaving

As a first example, consider the generalized SBC processes Constant A_{121}, A_{122} are defined as "$(r_{231} \bullet r_{232} \bullet STOP)$", "$(r_{233} \bullet r_{234} \bullet STOP)$". And the r_{231} (channel-based calling action under a tautology condition) prefix is defined as "$<C_{231}, CG, k_{231}>$", the r_{232} (channel-based calling action under a certain (tautology excluded) condition) prefix is defined as "**if** $cond_{232}$ **then** $<C_{231}, CD, k_{232}>$", the r_{233} (channel-based calling action under a tautology condition) prefix is defined as "$<C_{233}, CG, k_{233}>$", and the r_{234} (channel-based interaction under a tautology condition) prefix is defined as $<C_{234}, k_{234}, C_{235}>$.

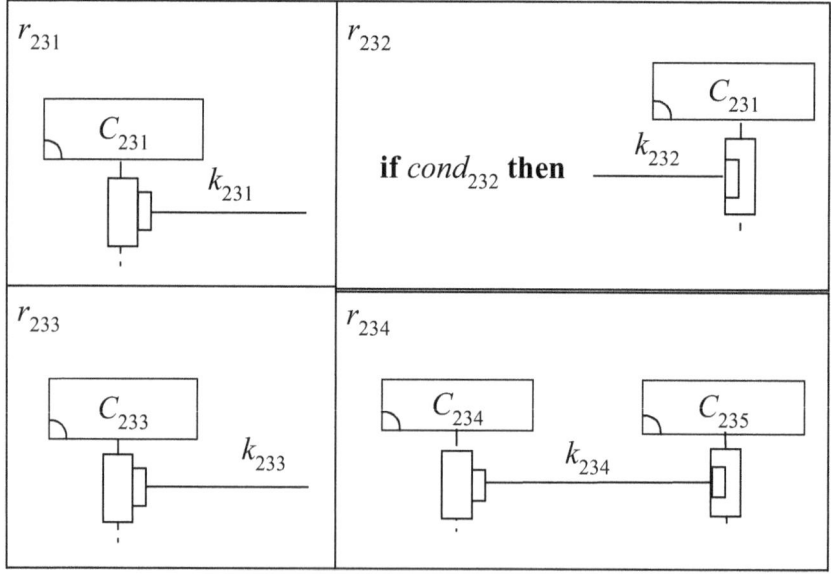

If we define A_{123} as "$A_{121}\|A_{122}$", the interleaving of processes A_{121} and A_{122}, then the following transition graph shows the semantics of process A_{123}.

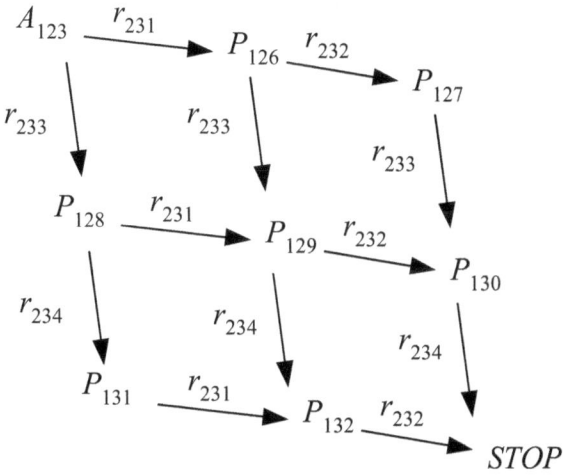

In the transition graph of the A_{123}'s generalized SBC process, processes A_{223}, P_{126}, P_{127}, P_{128}, P_{129}, P_{130}, P_{131} and P_{132} are defined as:

$$A_{123} \stackrel{\text{def}}{=} r_{231} \bullet P_{126} + r_{233} \bullet P_{128}$$

$$P_{126} \stackrel{\text{def}}{=} r_{232} \bullet P_{127} + r_{233} \bullet P_{129}$$

$$P_{127} \stackrel{\text{def}}{=} r_{233} \bullet P_{130}$$

$$P_{128} \stackrel{\text{def}}{=} r_{231} \bullet P_{129} + r_{234} \bullet P_{131}$$

$$P_{129} \stackrel{\text{def}}{=} r_{232} \bullet P_{130} + r_{234} \bullet P_{132}$$

$$P_{130} \stackrel{\text{def}}{=} r_{234} \bullet STOP$$

$$P_{131} \stackrel{\text{def}}{=} r_{231} \bullet P_{132}$$

$$P_{132} \stackrel{\text{def}}{=} r_{232} \bullet STOP$$

As a second example, consider the generalized SBC processes Constant A_{124}, A_{125}, A_{126} are defined as "$(r_{235} \bullet STOP)$", "$(r_{236} \bullet STOP)$", "$(r_{237} \bullet STOP)$". And the r_{235} (channel-based calling action under a tautology condition) prefix is defined as "$<C_{235}$, CD, $k_{235}>$", the r_{236} (channel-based calling action under a certain (tautology excluded) condition) prefix is defined as "if $cond_{236}$ then $<C_{236}$, CD, $k_{236}>$", and the r_{237} (channel-based calling action under a tautology condition) prefix is defined as "$<C_{237}$, CD, $k_{237}>$".

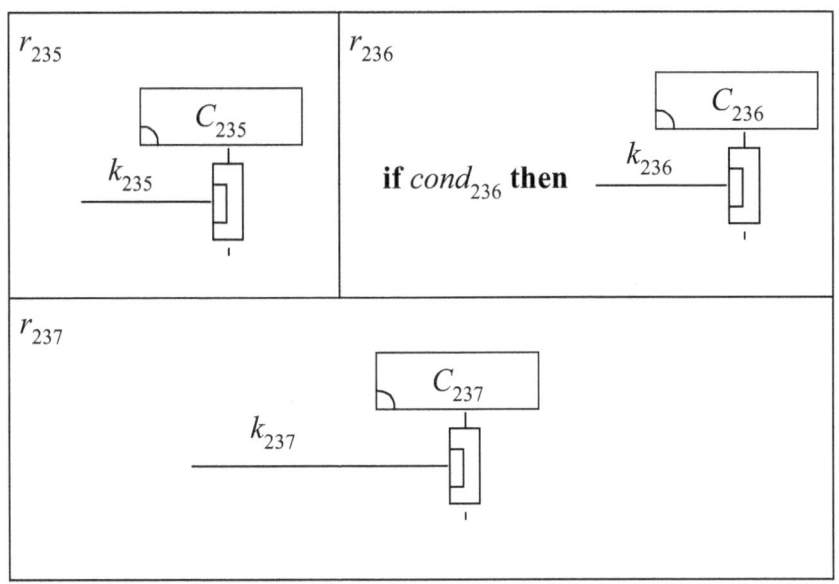

If we define A_{127} as "$A_{124}\|A_{125}\|A_{126}$", the interleaving of processes A_{124}, A_{125} and A_{126}, then the following transition graph shows the semantics of process A_{127}.

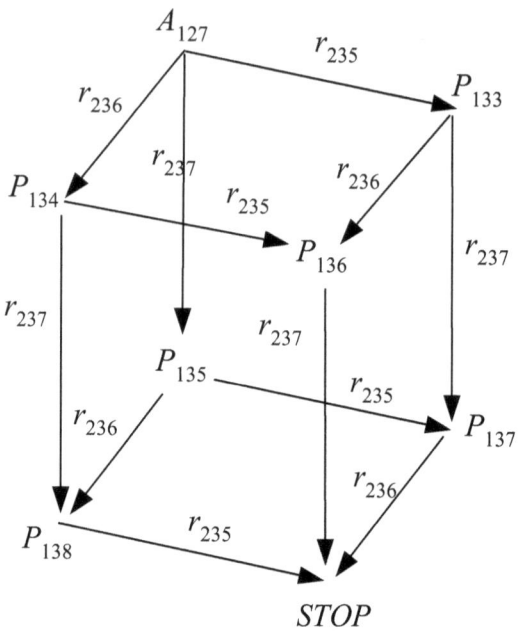

In the transition graph of the A_{127}'s generalized SBC process, processes A_{127}, P_{133}, P_{134}, P_{135}, P_{136}, P_{137} and P_{138} are defined as:

$$A_{127} \stackrel{def}{=} r_{235} \bullet P_{133} + r_{236} \bullet P_{134} + r_{237} \bullet P_{135}$$

$$P_{133} \stackrel{def}{=} r_{236} \bullet P_{136} + r_{237} \bullet P_{137}$$

$$P_{134} \stackrel{def}{=} r_{235} \bullet P_{136} + r_{237} \bullet P_{138}$$

$$P_{135} \stackrel{def}{=} r_{235} \bullet P_{137} + r_{236} \bullet P_{138}$$

$$P_{136} \stackrel{def}{=} r_{237} \bullet STOP$$

$$P_{137} \stackrel{def}{=} r_{236} \bullet STOP$$

$$P_{138} \stackrel{def}{=} r_{235} \bullet STOP$$

Operation-Based Interleaving

As a first example, consider the generalized SBC process Constant A_{128}, A_{129}, A_{130} are defined as "$\textbf{fix}(X_{128}=r_{238}\bullet X_{128})$", "$\textbf{fix}(X_{129}=r_{239}\bullet X_{129})$", "$\textbf{fix}(X_{130}=r_{240}\bullet r_{241}\bullet X_{130})$". And the r_{238} (operation-based called action under a tautology condition) prefix is defined as "$<$O_R, C_{238}, CD, $l_{238}>$", the r_{239} (operation-based called action under a certain (tautology excluded) condition) prefix is defined as "$\textbf{if } cond_{239} \textbf{ then } <$O_R, C_{239}, CD, $l_{239}>$", the r_{240} (operation-based called action under a tautology condition) prefix is defined as "$<$O_R, C_{240}, CD, $l_{240}>$", the r_{241} (operation-based called action under a certain (tautology excluded) condition) prefix is defined as "$\textbf{if } cond_{241} \textbf{ then } <$O_R, C_{241}, CD, $l_{241}>$".

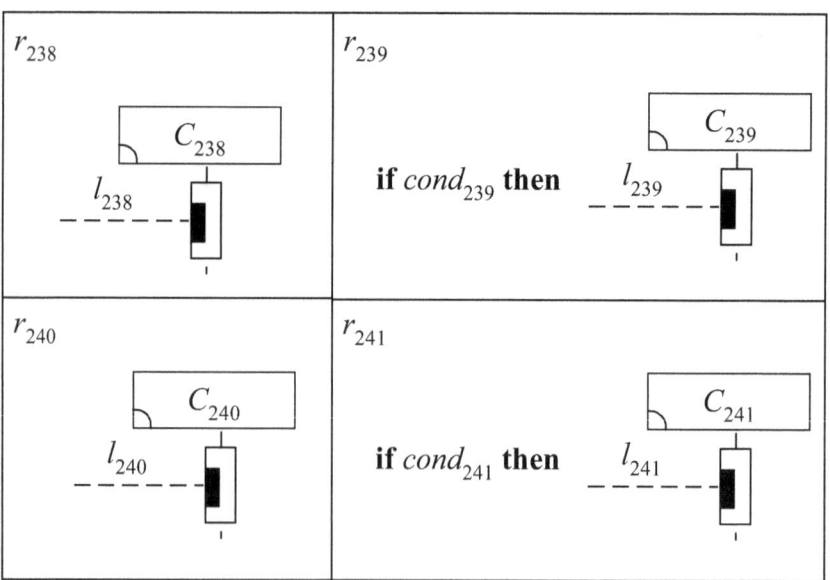

If we define A_{131} as "$A_{128}\|A_{129}\|A_{130}$", the interleaving of processes A_{128}, A_{129} and A_{130}, then the following transition graph shows the semantics of process A_{131}.

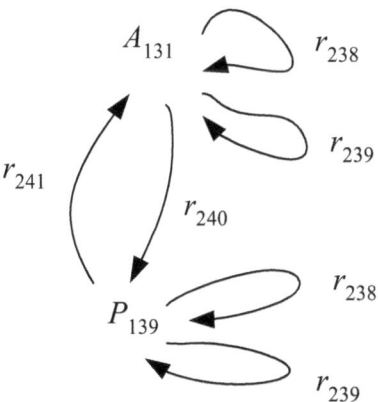

In the transition graph of the A_{131}'s generalized SBC process, processes A_{131} and P_{139} are defined as:

$$A_{131} \overset{\text{def}}{=\!=} r_{238} \bullet A_{131} + r_{239} \bullet A_{131} + r_{240} \bullet P_{139}$$

$$P_{139} \overset{\text{def}}{=\!=} r_{238} \bullet P_{139} + r_{239} \bullet P_{139} + r_{241} \bullet A_{131}$$

As a second example, consider the generalized SBC processes Constant A_{132}, A_{133} are defined as "$\mathbf{fix}(X_{132}=r_{242}\bullet r_{243}\bullet X_{132})$", "$\mathbf{fix}(X_{133}=r_{244}\bullet r_{245}\bullet X_{133})$". And the r_{242} (operation-based calling action under a tautology condition) prefix is defined as "$<O\_C, C_{242}, CG, l_{242}>$", the r_{243} (operation-based calling action under a certain (tautology excluded) condition) prefix is defined as "$\mathbf{if}\ cond_{243}\ \mathbf{then}\ <O\_C, C_{243}, CG, l_{243}>$", the r_{244} (operation-based calling action under a tautology condition) prefix is defined as "$<O\_C, C_{244}, CG, l_{244}>$", the r_{245} (operation-based calling action under a certain (tautology excluded) condition) prefix is defined as "$\mathbf{if}\ cond_{245}\ \mathbf{then}\ <O\_C, C_{245}, CG, l_{245}>$".

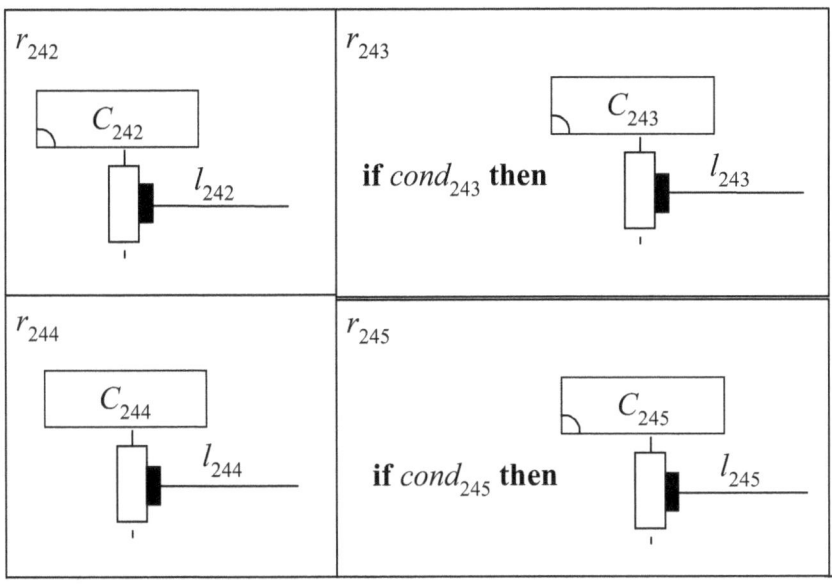

If we define A_{134} as "$A_{132}\|A_{133}$", the interleaving of processes A_{132} and A_{133}, then the following transition graph shows the semantics of process A_{134}.

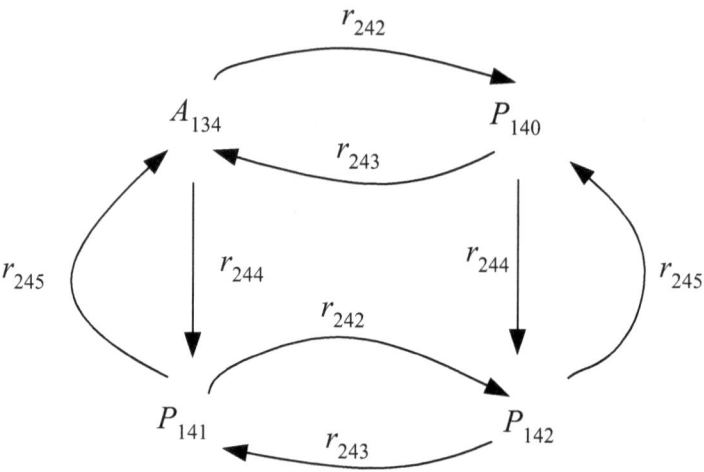

In the transition graph of the A_{134}'s generalized SBC process, processes A_{134}, P_{140}, P_{141} and P_{142} are defined as:

$$A_{134} \stackrel{\text{def}}{=\!=} r_{242} \bullet P_{140} + r_{244} \bullet P_{141}$$

$$P_{140} \stackrel{\text{def}}{=\!=} r_{243} \bullet A_{134} + r_{244} \bullet P_{142}$$

$$P_{141} \stackrel{\text{def}}{=\!=} r_{242} \bullet P_{142} + r_{245} \bullet A_{134}$$

$$P_{142} \stackrel{\text{def}}{=\!=} r_{243} \bullet P_{141} + r_{245} \bullet P_{140}$$

As a third example, consider the generalized SBC processes Constant A_{135}, A_{136} are defined as "**fix**$(X_{135}=r_{246} \bullet r_{247} \bullet X_{135})$", "**fix**$(X_{136}=r_{248} \bullet r_{249} \bullet r_{250} \bullet X_{136})$". And the r_{246} (operation-based calling action under a tautology condition) prefix is defined as "<O_R, C_{246}, CG, l_{246}>", the r_{247} (operation-based calling action under a certain (tautology excluded) condition) prefix is defined as "**if** $cond_{247}$ **then** <O_R, C_{247}, CG, l_{247}>", the r_{248} (operation-based calling action under a tautology condition) prefix is defined as "<O_R, C_{248}, CG, l_{248}>", the r_{249} (operation-based calling action under a tautology condition) prefix is defined as "<O_R, C_{249}, CG, l_{249}>", the r_{250} (operation-based calling action under a tautology condition) prefix is defined as "<O_R, C_{250}, CG, l_{250}>",

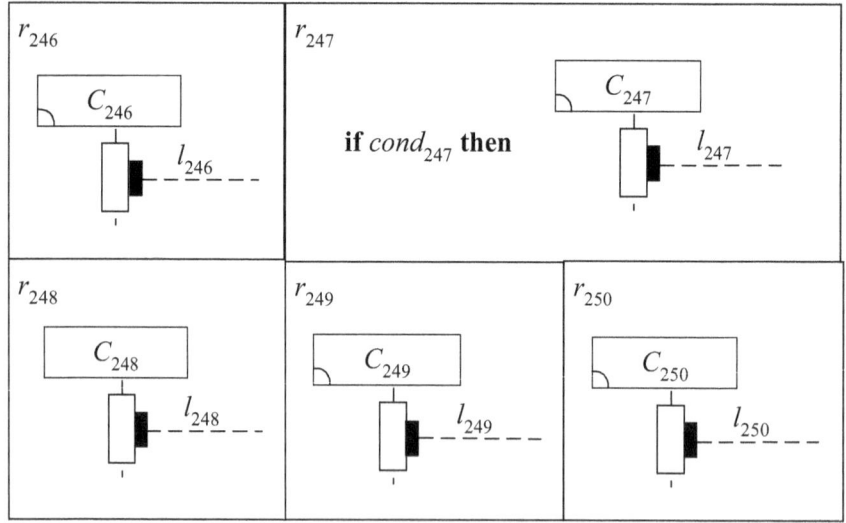

If we define A_{137} as "$A_{135}\|A_{136}$", the interleaving of processes A_{135} and A_{136}, then the following transition graph shows the semantics of process A_{137}.

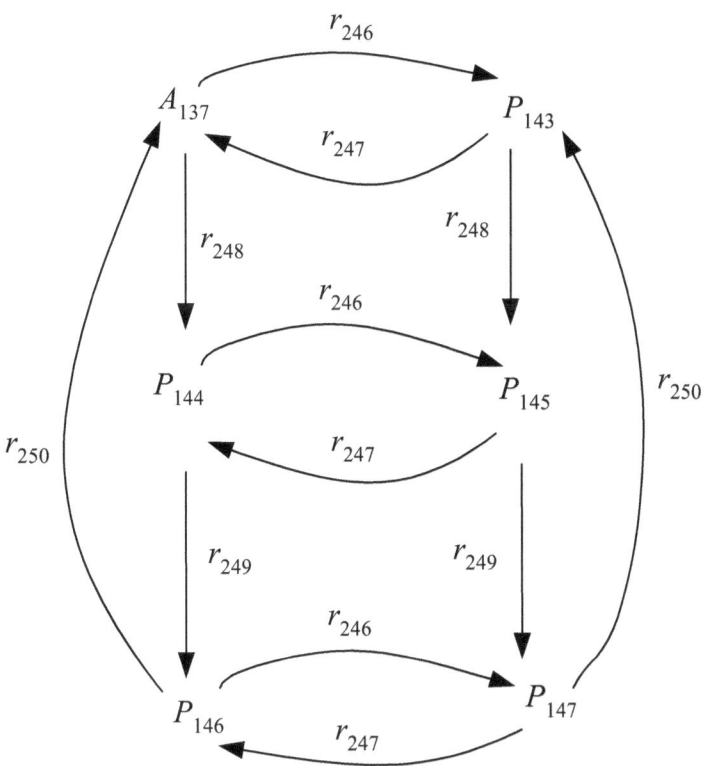

In the transition graph of the A_{137}'s generalized SBC process, processes A_{137}, P_{143}, P_{144}, P_{145}, P_{146} and P_{147} are defined as:

$$A_{137} \overset{\text{def}}{=} r_{246} \bullet P_{143} + r_{248} \bullet P_{144}$$

$$P_{143} \overset{\text{def}}{=} r_{247} \bullet A_{137} + r_{248} \bullet P_{145}$$

$$P_{144} \overset{\text{def}}{=} r_{246} \bullet P_{145} + r_{249} \bullet P_{146}$$

$$P_{145} \overset{\text{def}}{=} r_{247} \bullet P_{144} + r_{249} \bullet P_{147}$$

$$P_{146} \overset{\text{def}}{=} r_{246} \bullet P_{147} + r_{250} \bullet A_{137}$$

$$P_{147} \overset{\text{def}}{=} r_{247} \bullet P_{146} + r_{250} \bullet P_{143}$$

PART IX: EXAMPLES OF RESTRICTED COMPOSITION

Restricted Composition

Restricted composition (or expansion law) [Miln89, Miln99] is also a kind of parallel composition. In general, transition rules Parallel$_3$, Parallel$_4$, Parallel$_5$, Paralle$_6$, Parallel$_7$, or Parallel$_8$ will be applied for the purpose of restricted composition.

Very often a concurrent system is naturally expressed as a restricted composition, i.e. in the form $(P_1\|P_2\|...\|P_n)\backslash H$. Here we demonstrate several examples of both channel-based and operation-based restricted composition.

Channel-Based Restricted Composition

As a first example, consider the generalized SBC process Constant A_{301} is defined as "$r_{301} \bullet STOP + r_{302} \bullet STOP$" and the r_{301} (channel-based calling action under a certain (tautology excluded) condition) prefix is defined as "if $cond_{301}$ **then** $<C_{301}$, CG, $k_{301}>$" and the r_{302} (channel-based calling action under a certain (tautology excluded) condition) prefix is defined as "if $cond_{302}$ **then** $<C_{302}$, CG, $k_{302}>$". Also consider the generalized SBC process Constant A_{302} is defined as "$r_{303} \bullet STOP$" and the r_{303} (channel-based called action under a certain (tautology excluded) condition) prefix is defined as "if $cond_{303}$ **then** $<C_{303}$, CD, $k_{301}>$".

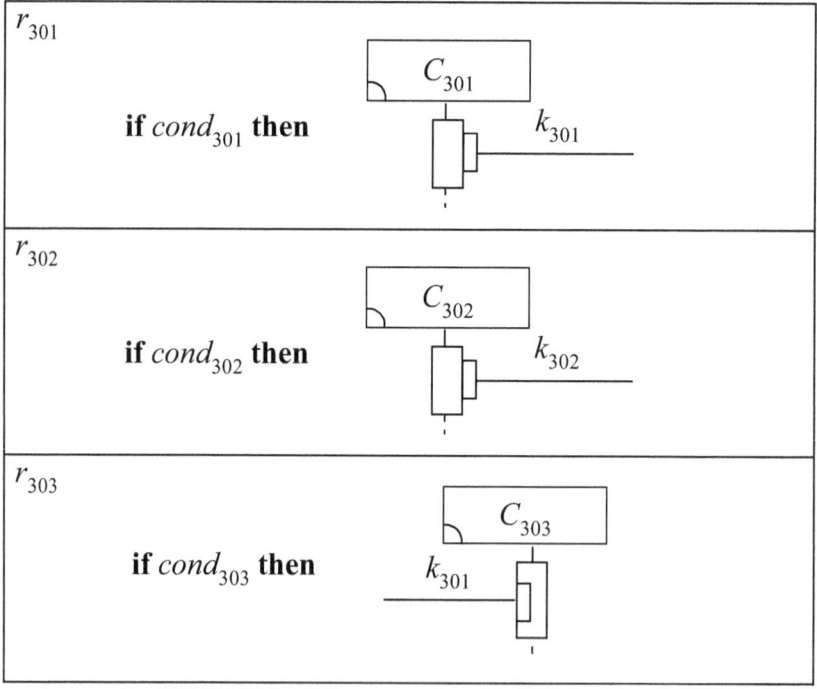

We define the generalized SBC process Constant A_{303} as the parallel composition of A_{301} and A_{302}, i.e. $(A_{301}\|A_{302})$ and the r_{304} (channel-based interaction under a certain (tautology excluded) condition) prefix is defined as "if $cond_{301}\&cond_{303}$ then $<C_{301}, k_{301}, C_{303}>$".

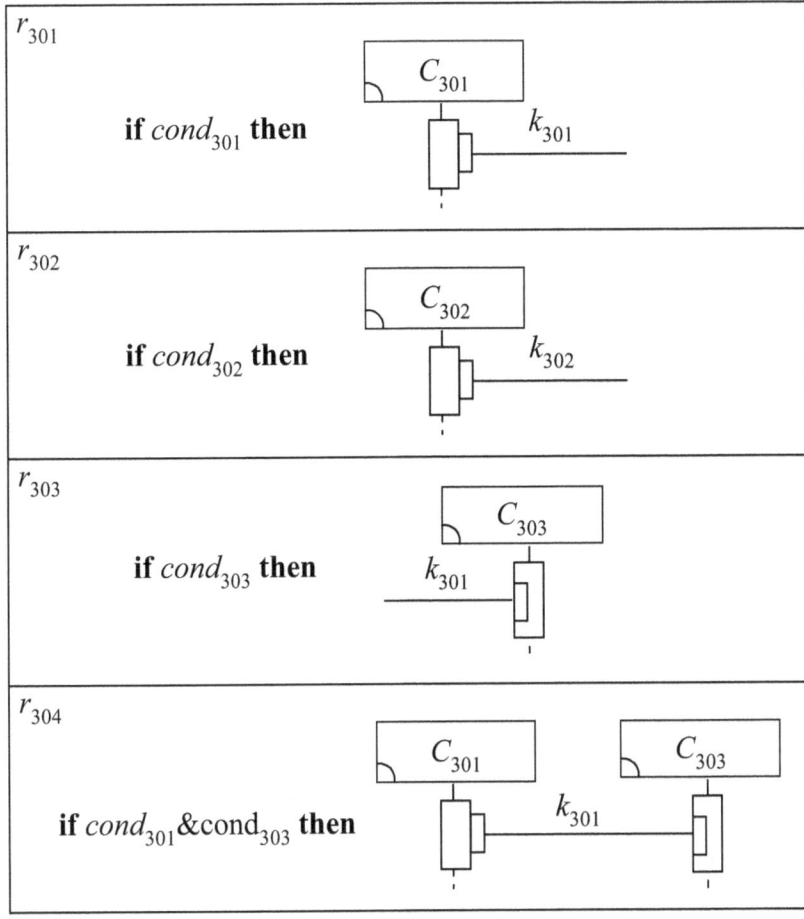

The following transition graph shows the semantics of process A_{303}.

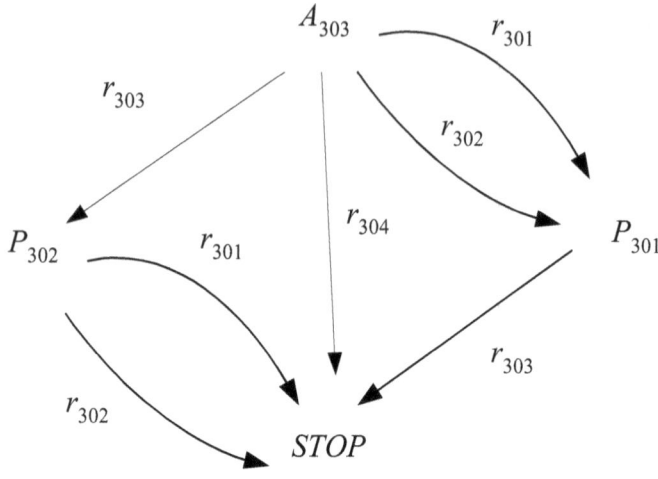

In the transition graph of the A_{303}'s generalized SBC process, processes A_{303}, P_{301} and P_{302} are defined as:

$$A_{303} \overset{\text{def}}{=\joinrel=} r_{301} \bullet P_{301} + r_{302} \bullet P_{301} + r_{303} \bullet P_{302} + r_{304} \bullet STOP$$

$$P_{301} \overset{\text{def}}{=\joinrel=} r_{303} \bullet STOP$$

$$P_{302} \overset{\text{def}}{=\joinrel=} r_{301} \bullet STOP + r_{302} \bullet STOP$$

We define the generalized SBC process Constant A_{304} as the restricted composition of A_{301} and A_{302}, i.e. $(A_{301}\|A_{302})\backslash\{r_{301}, r_{303}\}$ and the r_{302} (channel-based calling action under a certain (tautology excluded) condition) prefix is defined as "**if** $cond_{302}$ **then** $<C_{302}$, CG, $k_{302}>$" and the r_{304} (channel-based interaction under a certain (tautology excluded) condition) prefix is defined as "**if** $cond_{301}\&cond_{303}$ **then** $<C_{301}, k_{301}, C_{303}>$".

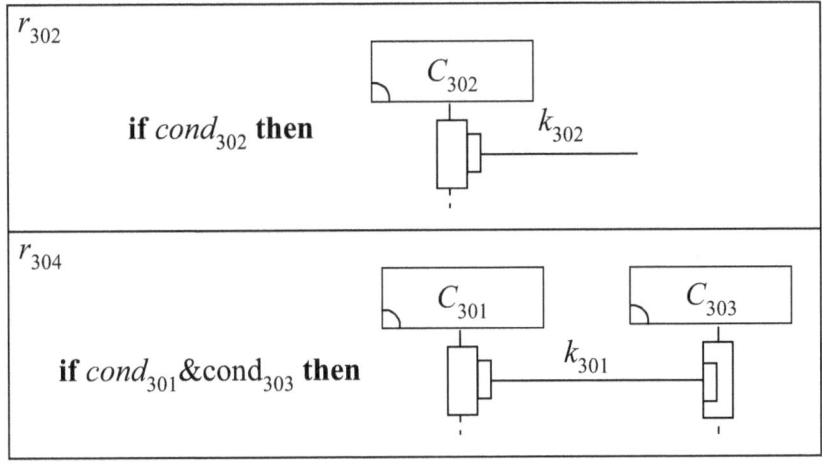

The following transition graph shows the semantics of process A_{304}.

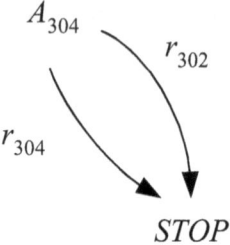

In the transition graph of the A_{304}'s generalized SBC process, processes A_{304} is defined as:

$$A_{304} \overset{\text{def}}{=\joinrel=} r_{302} \bullet STOP + r_{304} \bullet STOP$$

As a second example, consider the generalized SBC process Constant A_{305} is defined as "$\mathbf{fix}(X_{305}=r_{305}\bullet r_{306}\bullet X_{305})$", the r_{305} (channel-based called action under a tautology condition) prefix is defined as "$<C_{305}, CD, k_{305}>$" and the r_{306} (channel-based calling action under a certain (tautology excluded) condition) prefix is defined as "$\mathbf{if}\ cond_{306}\ \mathbf{then}\ <C_{306}, CG, k_{306}>$". Also consider the generalized SBC process Constant A_{306} is defined as "$\mathbf{fix}(X_{306}=r_{307}\bullet r_{308}\bullet X_{306})$", the r_{307} (channel-based called action under a tautology condition) prefix is defined as "$<C_{307}, CD, k_{306}>$" and the r_{308} (channel-based calling action under a tautology condition) prefix is defined as "$<C_{308}, CG, k_{308}>$".

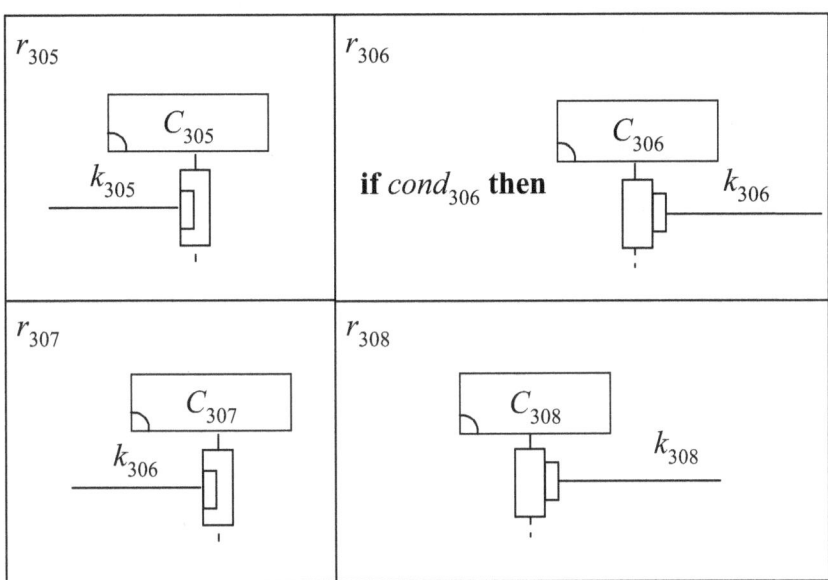

We define the generalized SBC process Constant A_{307} as the parallel composition of A_{305} and A_{306}, i.e. $(A_{305}\|A_{306})$ and the r_{309} (channel-based interaction under a certain (tautology excluded) condition) prefix is defined as "**if** $cond_{306}$ **then** $<C_{306}, k_{306}, C_{307}>$".

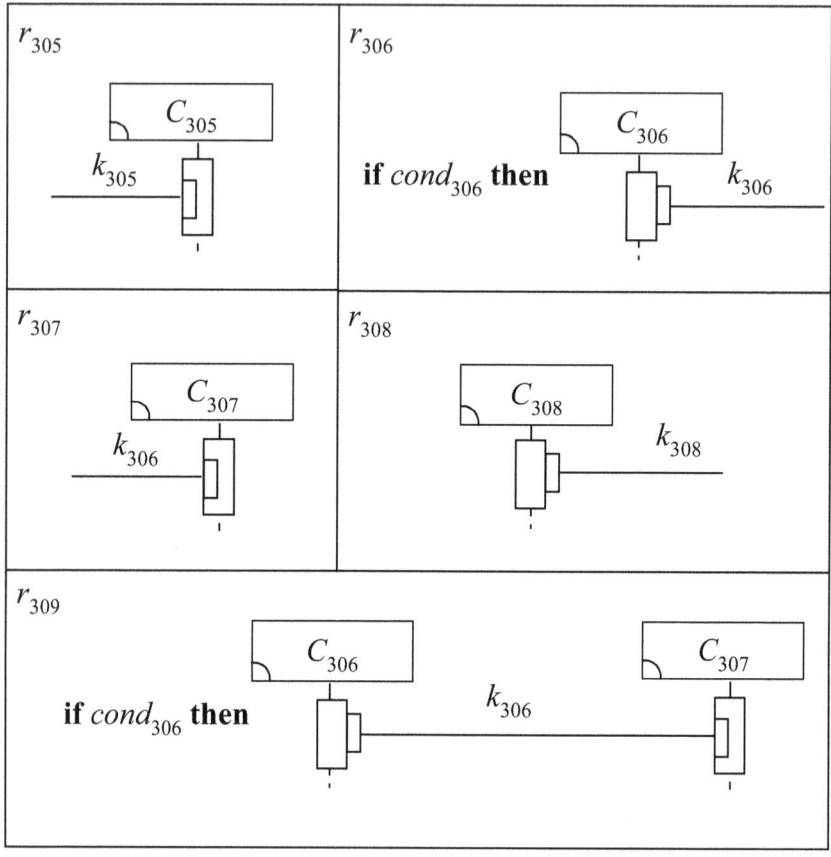

The following transition graph shows the semantics of process A_{307}.

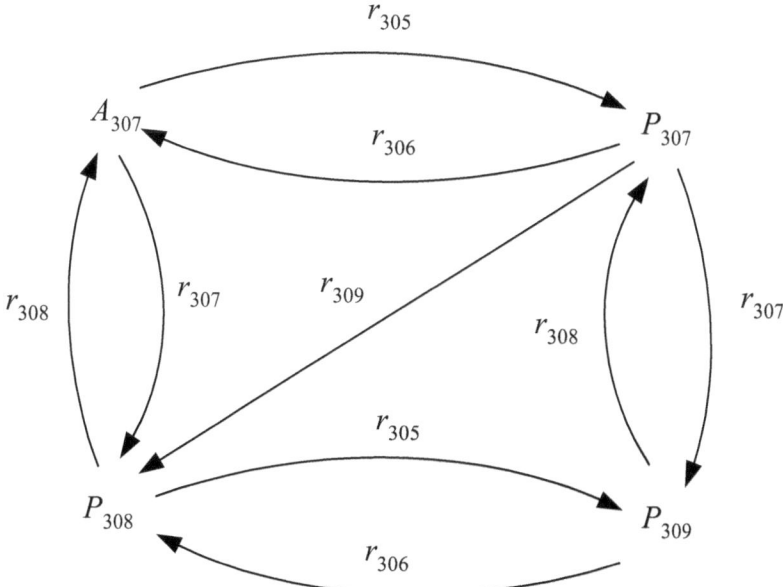

In the transition graph of the A_{307}'s generalized SBC process, processes A_{307}, P_{307}, P_{308} and P_{309} are defined as:

$$A_{307} \stackrel{\text{def}}{=\joinrel=} r_{305} \bullet P_{307} + r_{307} \bullet P_{308}$$

$$P_{307} \stackrel{\text{def}}{=\joinrel=} r_{307} \bullet P_{309} + r_{306} \bullet A_{307} + r_{309} \bullet P_{308}$$

$$P_{308} \stackrel{\text{def}}{=\joinrel=} r_{305} \bullet P_{309} + r_{308} \bullet A_{307}$$

$$P_{309} \stackrel{\text{def}}{=\joinrel=} r_{308} \bullet P_{307} + r_{306} \bullet P_{308}$$

We define the generalized SBC process Constant A_{308} as the restricted composition of A_{305} and A_{306}, i.e. $(A_{305}\|A_{306})\backslash\{r_{306}, r_{307}\}$ and the r_{309} (channel-based interaction under a certain (tautology excluded) condition) prefix is defined as "**if** $cond_{306}$ **then** $<C_{306}, k_{306}, C_{307}>$".

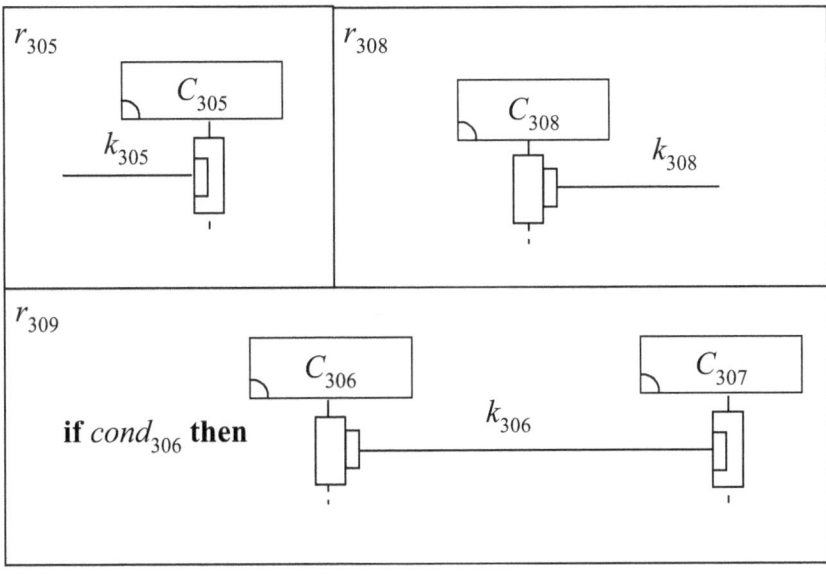

The following transition graph shows the semantics of process A_{308}.

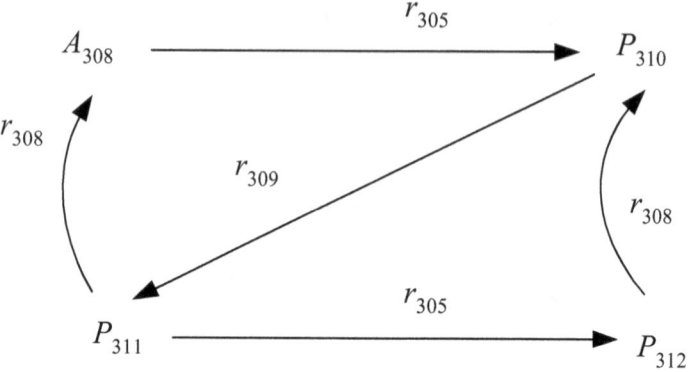

In the transition graph of the A_{308}'s generalized SBC process, processes A_{308}, P_{310}, P_{311} and P_{312} are defined as:

$$A_{308} \overset{\text{def}}{=\!=} r_{305} \bullet P_{310}$$

$$P_{310} \overset{\text{def}}{=\!=} r_{309} \bullet P_{311}$$

$$P_{311} \overset{\text{def}}{=\!=} r_{305} \bullet P_{312} + r_{308} \bullet A_{308}$$

$$P_{312} \overset{\text{def}}{=\!=} r_{308} \bullet P_{310}$$

Operation-Based Restricted Composition

As a first example, consider the generalized SBC process Constant A_{311} is defined as "$r_{311} \bullet STOP$" and the r_{311} (operation-based calling action under a tautology condition) prefix is defined as "<O_R, C_{311}, CG, l_{311}>". Also consider the generalized SBC process Constant A_{312} is defined as "$r_{312} \bullet STOP$" and the r_{312} (operation-based called action under a certain (tautology excluded) condition) prefix is defined as "**if** $cond_{312}$ **then** <O_R, C_{312}, CD, l_{311}>.

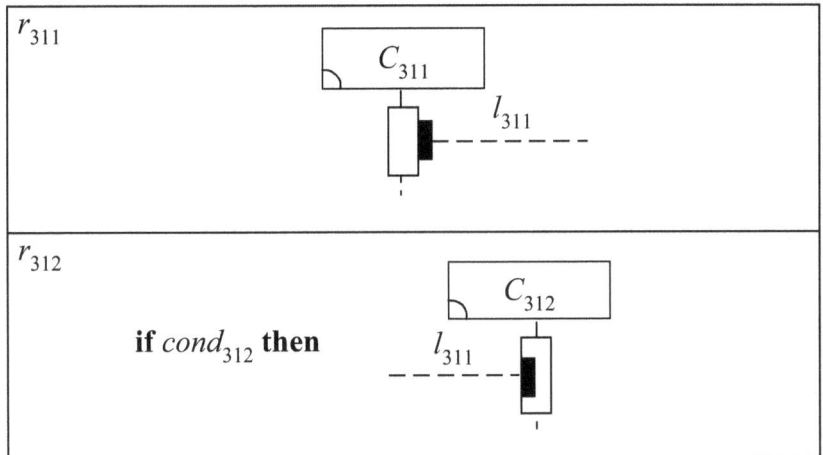

We define the generalized SBC process Constant A_{313} as the parallel composition of A_{311} and A_{312}, i.e. $(A_{311}\|A_{312})$ and the r_{313} (operation-based interaction under a certain (tautology excluded) condition) prefix is defined as "**if** $cond_{312}$ **then** $<O\_R, C_{311}, l_{311}, C_{312}>$".

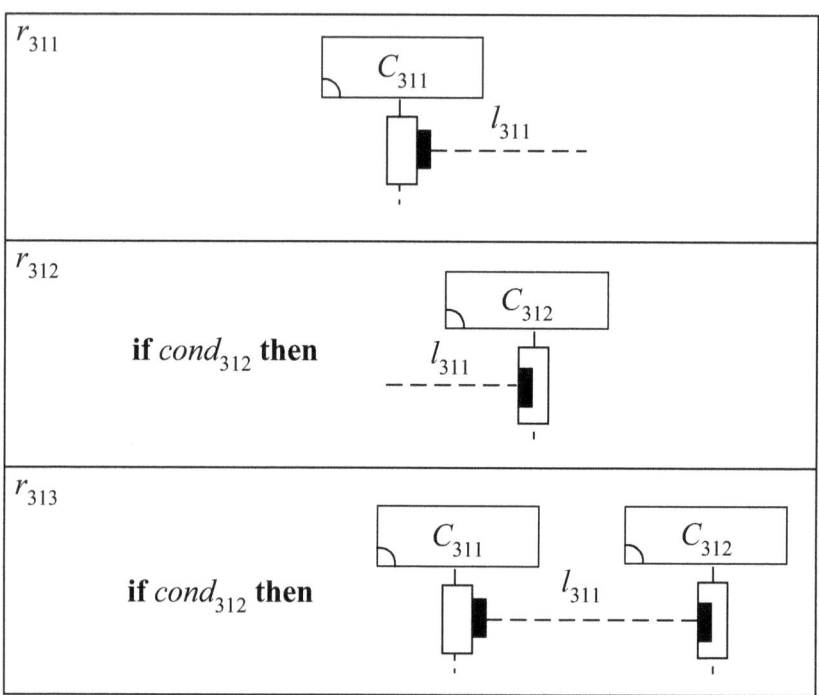

The following transition graph shows the semantics of process A_{313}.

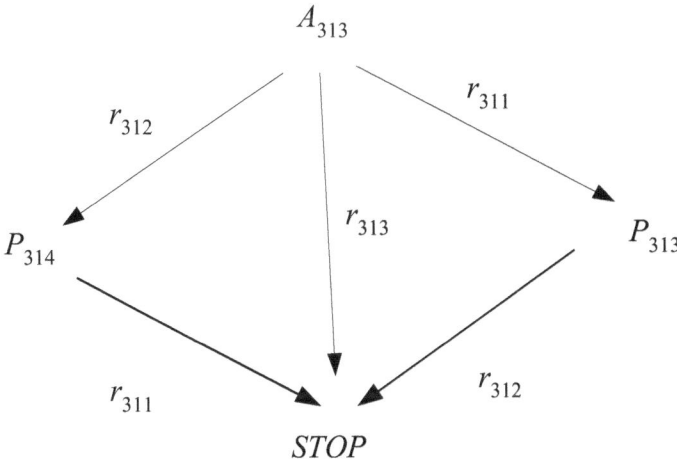

In the transition graph of the A_{313}'s generalized SBC process, processes A_{313}, P_{313} and P_{314} are defined as:

$$A_{313} \stackrel{\text{def}}{=\!=} r_{311} \bullet P_{313} + r_{312} \bullet P_{314} + r_{311} \bullet STOP$$

$$P_{313} \stackrel{\text{def}}{=\!=} r_{312} \bullet STOP$$

$$P_{314} \stackrel{\text{def}}{=\!=} r_{313} \bullet STOP$$

We define the generalized SBC process Constant A_{314} as the restricted composition of A_{311} and A_{312}, i.e. $(A_{311}\|A_{312})\backslash\{r_{311},\ r_{312}\}$ and the r_{313} (operation-based interaction under a certain (tautology excluded) condition) prefix is defined as "**if** $cond_{312}$ **then** <O_R, C_{311}, l_{311}, C_{312}>".

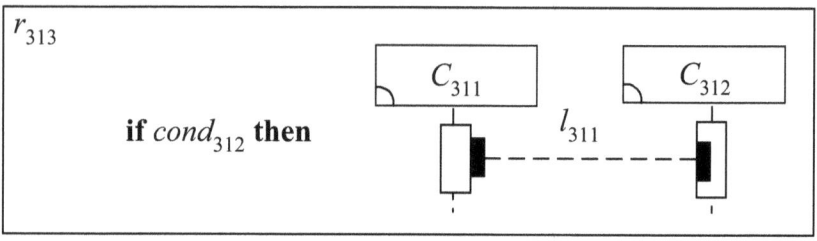

The following transition graph shows the semantics of process A_{314}.

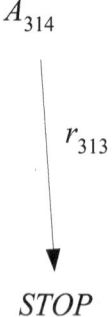

In the transition graph of the A_{314}'s generalized SBC process, processes A_{314} is defined as:

$$A_{314} \stackrel{\text{def}}{=\!=} r_{313} \bullet STOP$$

As a second example, consider the generalized SBC process Constant A_{321} is defined as "$\mathbf{fix}(X_{321}=r_{321} \bullet r_{322} \bullet X_{321})$", the r_{321} (operation-based called action under a tautology condition) prefix is defined as "$<O\_C, C_{321}, CD, l_{321}>$" and the r_{322} (operation-based calling action under a certain (tautology excluded) condition) prefix is defined as "$\mathbf{if}\ cond_{322}\ \mathbf{then}\ <O\_C, C_{322}, CG, l_{322}>$". Also consider the generalized SBC process Constant A_{322} is defined as "$\mathbf{fix}(X_{322}=r_{323} \bullet r_{324} \bullet X_{322})$", the r_{323} (operation-based called action under a certain (tautology excluded) condition) prefix is defined as "$\mathbf{if}\ cond_{323}\ \mathbf{then}\ <O\_C, C_{323}, CD, l_{322}>$" and the r_{324} (operation-based calling action under a tautology condition) prefix is defined as "$<O\_C, C_{324}, CG, l_{324}>$".

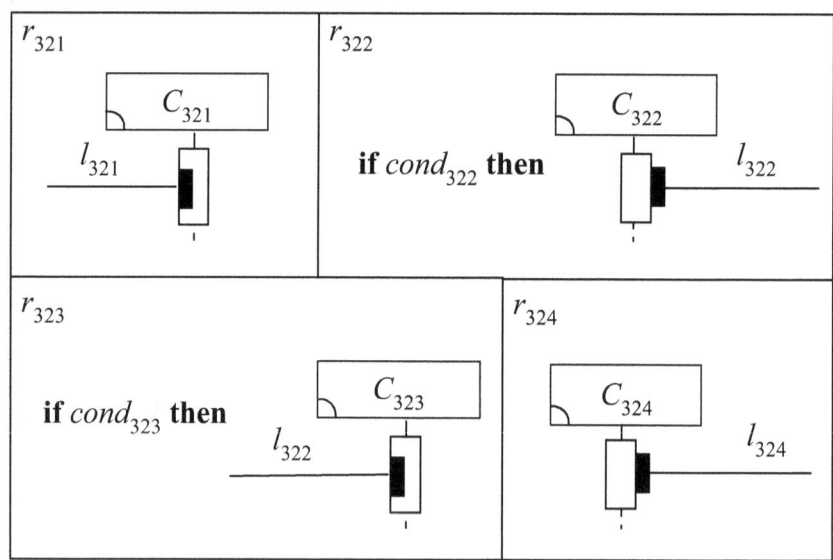

We define the generalized SBC process Constant A_{323} as the parallel composition of A_{321} and A_{322}, i.e. $(A_{321}\|A_{322})$ and the r_{325} (operation-based interaction under a certain (tautology excluded) condition) prefix is defined as "**if** $cond_{322}\&cond_{323}$ **then** $<O\_C,$ $C_{322}, l_{322}, C_{323}>$".

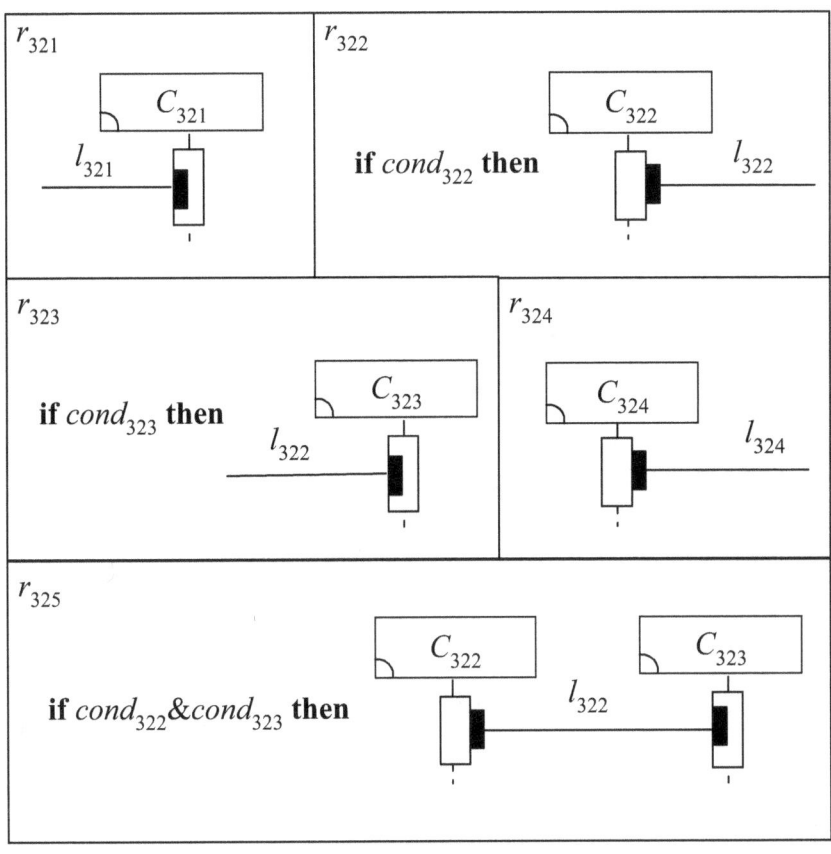

The following transition graph shows the semantics of process A_{323}.

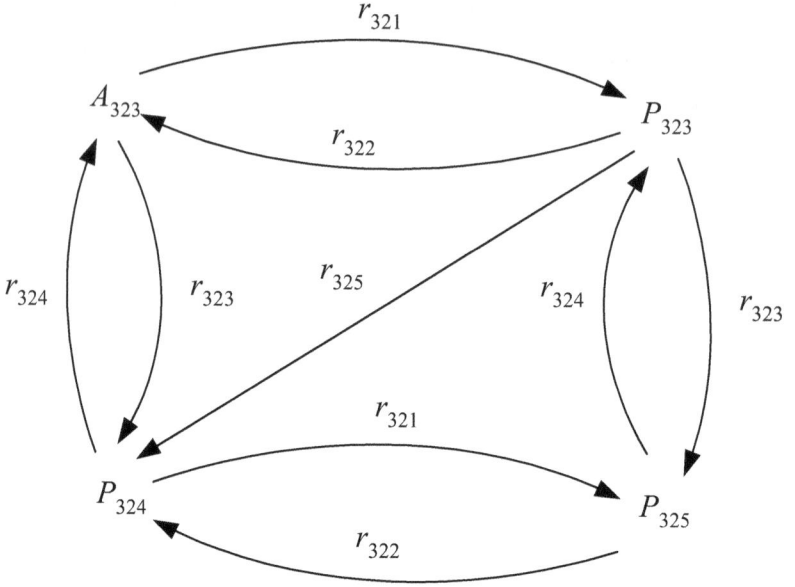

In the transition graph of the A_{323}'s generalized SBC process, processes A_{323}, P_{323}, P_{324} and P_{325} are defined as:

$$A_{323} \stackrel{\text{def}}{=\!=} r_{321} \bullet P_{323} + r_{323} \bullet P_{324}$$

$$P_{323} \stackrel{\text{def}}{=\!=} r_{323} \bullet P_{325} + r_{322} \bullet A_{323} + r_{325} \bullet P_{324}$$

$$P_{324} \stackrel{\text{def}}{=\!=} r_{321} \bullet P_{325} + r_{324} \bullet A_{323}$$

$$P_{325} \stackrel{\text{def}}{=\!=} r_{324} \bullet P_{323} + r_{322} \bullet P_{324}$$

We define the generalized SBC process Constant A_{324} as the restricted composition of A_{321} and A_{322}, i.e. $(A_{321}\|A_{322})\backslash\{r_{322}, r_{323}\}$ and the r_{325} (operation-based interaction under a certain (tautology excluded) condition) prefix is defined as "**if** $cond_{322}\&cond_{323}$ **then** $<O\_C, C_{322}, l_{322}, C_{323}>$".

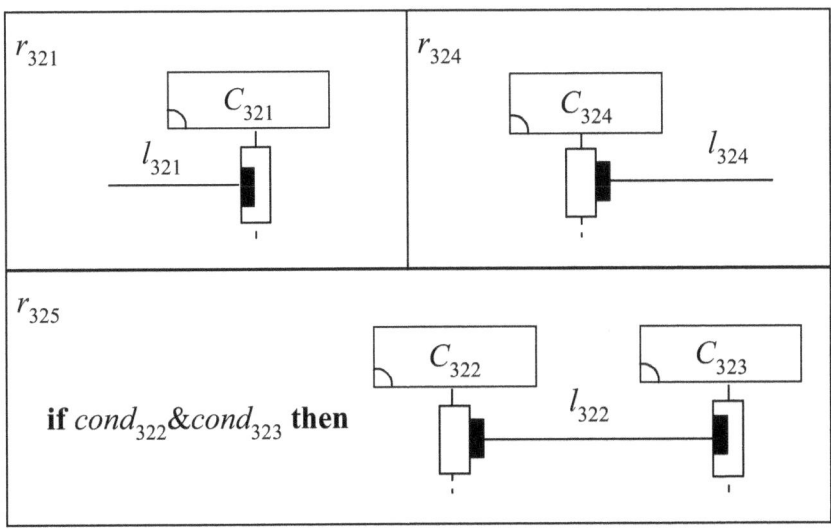

The following transition graph shows the semantics of process A_{324}.

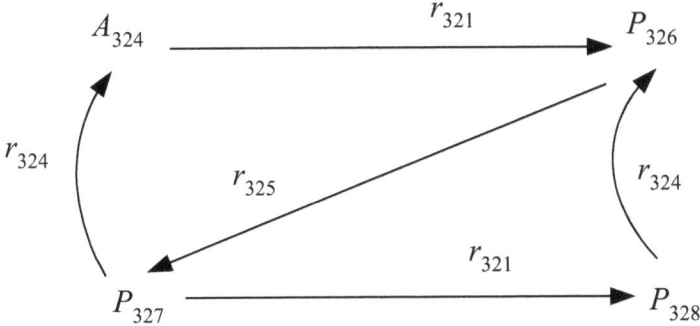

In the transition graph of the A_{324}'s generalized SBC process, processes A_{324}, P_{326}, P_{327} and P_{328} are defined as:

$$A_{324} \stackrel{\text{def}}{=\!\!=} r_{321} \bullet P_{326}$$

$$P_{326} \stackrel{\text{def}}{=\!\!=} r_{325} \bullet P_{327}$$

$$P_{327} \stackrel{\text{def}}{=\!\!=} r_{321} \bullet P_{328} + r_{324} \bullet A_{324}$$

$$P_{328} \stackrel{\text{def}}{=\!\!=} r_{324} \bullet P_{326}$$

PART X: EXAMPLES OF STRUCTURAL COMPOSITION

Structural Composition

Structural decomposition and composition may help us better understand a system. For example, the "tree" system is composed of "root" and "stem" as shown below.

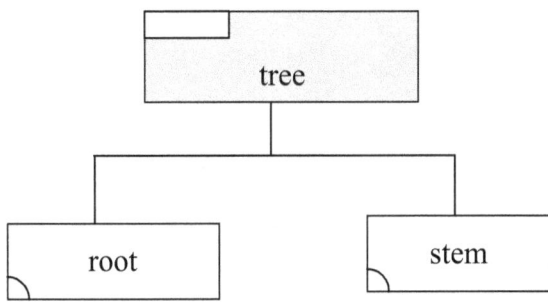

Generalized SBC process algebra uses the renaming function to accomplish the meaning of structural composition. For each renaming function f, the renaming combinatory $[f]$, postfixed to a process or a condition action/interaction, has the effect of renaming the components (of the process or condition action/interaction) as dictated by f. We often write $C'_1/C_1,\ldots, C'_n/C_n$ for the renaming function for which $f(C_i) = C'_i$ for $i = 1,\ldots, n$.

After renaming, an interaction under a tautology condition will be internalized to be an internal interaction (i.e. λ) if it describes that a component interacts with itself.

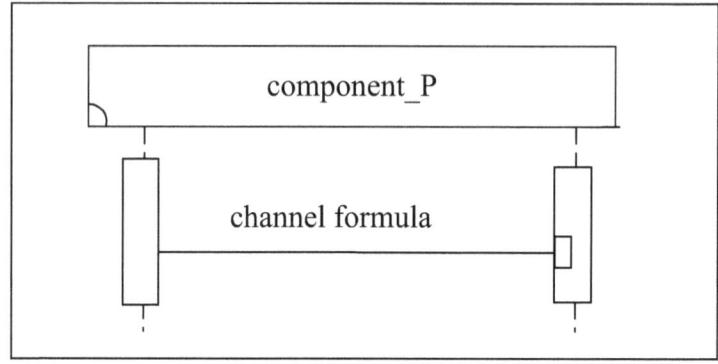

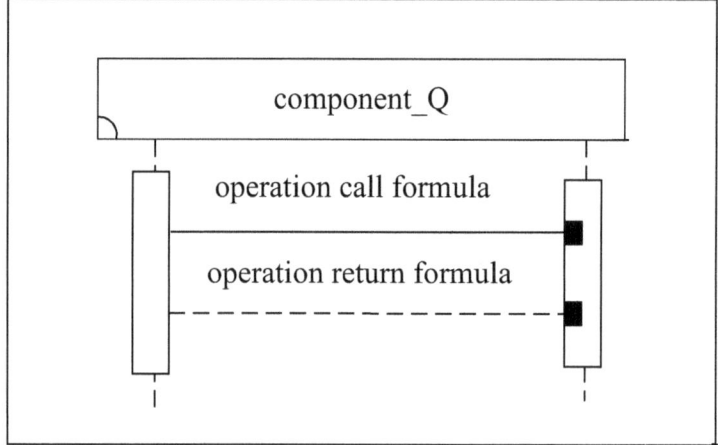

Channel-Based Structural Composition

As a first example, consider the generalized SBC process Constant A_{401} is defined as "$r_{401} \bullet r_{402} \bullet r_{403} \bullet STOP[C_{406}/C_{401}, C_{406}/C_{402}]$" and the r_{401} (channel-based interaction under a tautology condition) prefix is defined as "$<C_{401}, k_{401}, C_{402}>$", the r_{402} (channel-based interaction under a tautology condition) prefix is defined as "$<C_{403}, k_{402}, C_{404}>$", the r_{403} (channel-based calling action under a certain (tautology excluded) condition) prefix are defined as "**if** $cond_{403}$ **then** $<C_{405}, CG, k_{403}>$".

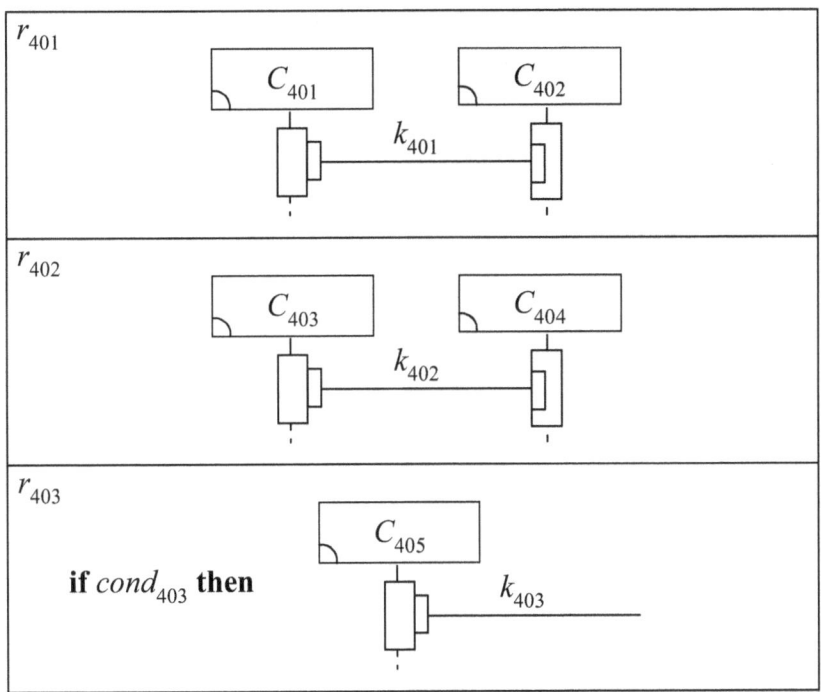

Structural composition of "C_{401}" and "C_{402}" into "C_{406}" means to rename both the "C_{401}" and "C_{402}" component to the "C_{406}" component.

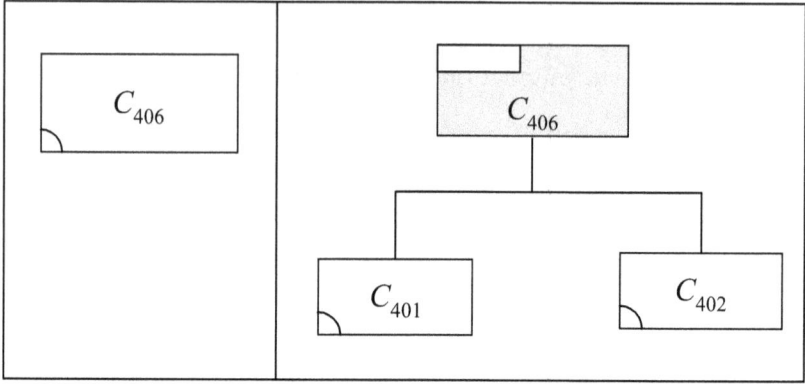

After structurally composing of "C_{401}" and "C_{402}" into "C_{406}", the r_{401} prefix becomes the r_{404} (channel-based interaction under a tautology condition) prefix represented by $<C_{406}, k_{401}, C_{406}>$ which stands for an internal interaction (i.e. λ) because it describes the "C_{406}" component interacts with itself.

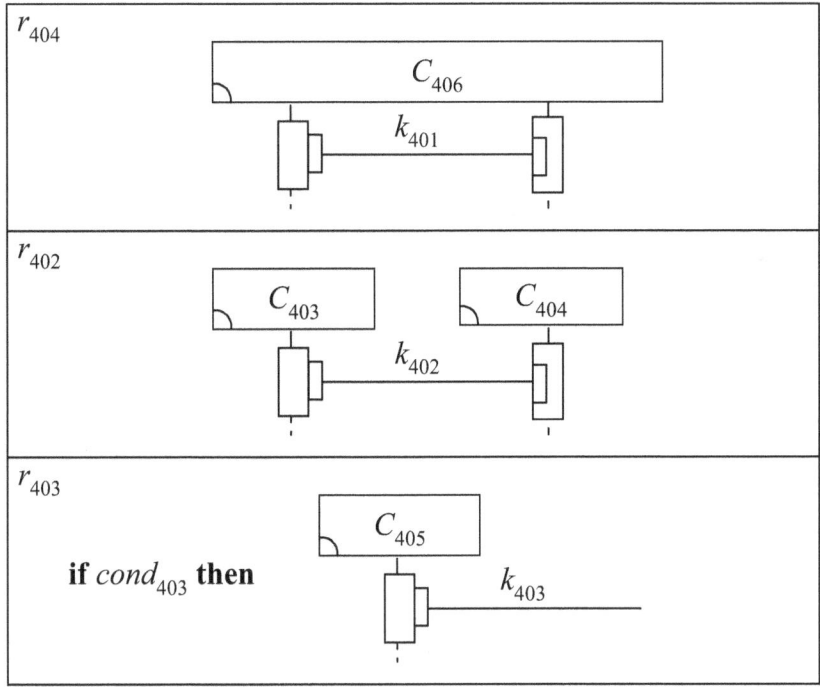

The following transition graph shows the semantics of process A_{401}.

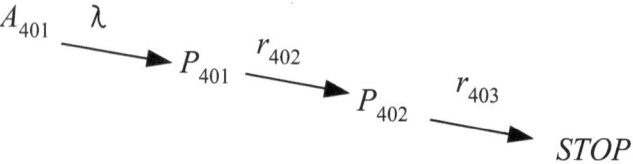

In the transition graph of the A_{401}'s generalized SBC process, processes A_{401}, P_{401} and P_{402} are defined as:

$$A_{401} \stackrel{\text{def}}{=} \lambda \bullet P_{401}$$

$$P_{401} \stackrel{\text{def}}{=} r_{402} \bullet P_{402}$$

$$P_{402} \stackrel{\text{def}}{=} r_{403} \bullet STOP$$

As a second example, consider the generalized SBC process Constant A_{411} is defined as "$r_{411} \bullet r_{412} \bullet STOP[C_{415}/C_{413}, C_{415}/C_{414}]$" and the r_{411} (channel-based interaction under a tautology condition) prefix is defined as "$<C_{411}, k_{411}, C_{412}>$", the r_{412} (channel-based interaction under a certain (tautology excluded) condition) prefix is defined as "**if** $cond_{412}$ **then** $<C_{413}, k_{412}, C_{414}>$".

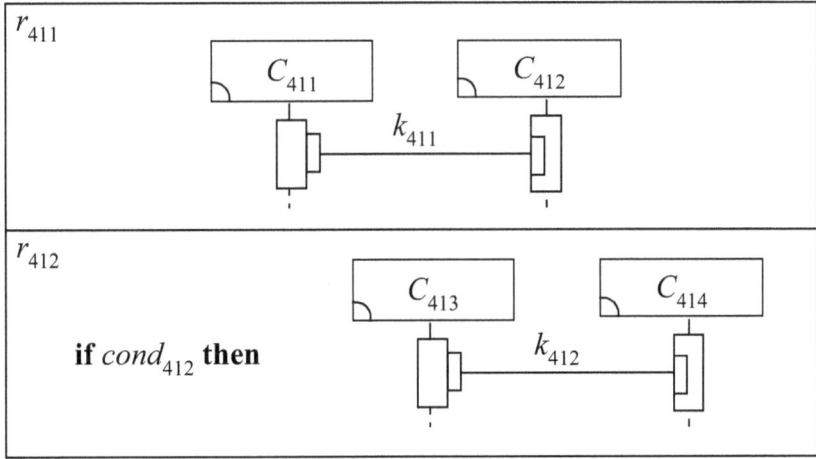

182

Structural composition of "C_{413}" and "C_{414}" into "C_{415}" means to rename both the "C_{413}" and "C_{414}" component to the "C_{415}" component.

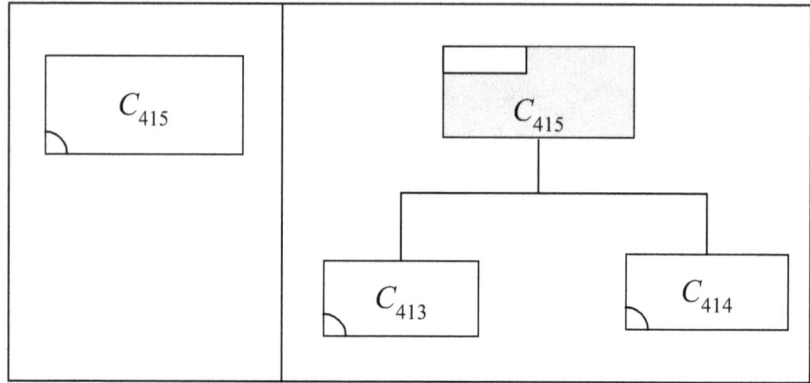

After structurally composing of "C_{413}" and "C_{414}" into "C_{415}", the r_{412} prefix becomes the r_{413} (channel-based interaction under a certain (tautology excluded) condition) prefix represented by "**if** $cond_{412}$ **then** $<C_{415}, k_{412}, C_{415}>$" which stands for an internal interaction (i.e. λ) under a certain (tautology excluded) condition. (Be noted that an internal interaction under a certain (tautology excluded) condition is not equal to an internal interaction.)

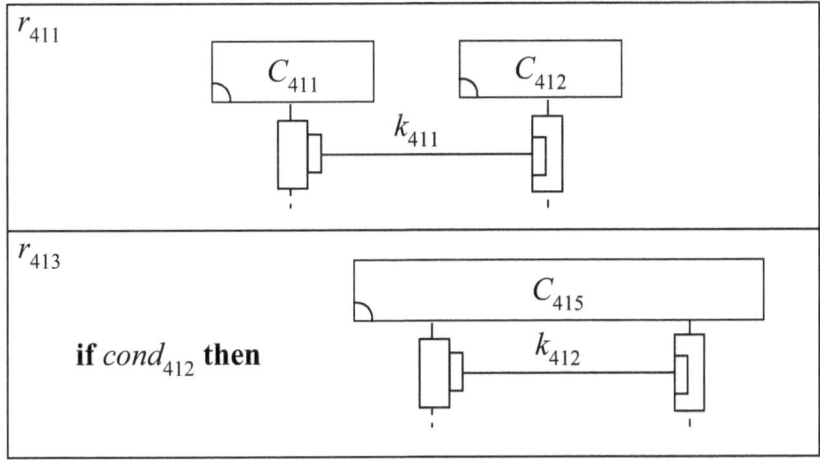

The following transition graph shows the semantics of process A_{411}.

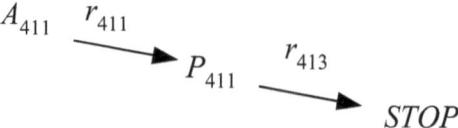

In the transition graph of the A_{411}'s generalized SBC process, processes A_{411} and P_{411} are defined as:

$$A_{411} \overset{\text{def}}{=\joinrel=} r_{411} \bullet P_{411}$$

$$P_{411} \overset{\text{def}}{=\joinrel=} r_{413} \bullet STOP$$

As a third example, consider the generalized SBC process Constant A_{421} is defined as "**fix**$(X_{421}=r_{421} \bullet X_{421}+r_{422} \bullet r_{423} \bullet X_{421})[C_{425}/C_{422}, C_{425}/C_{423}]$" and the r_{421} (channel-based called action under a tautology condition) prefix is defined as "$<C_{421}, CD, k_{421}>$", the r_{422}(channel-based interaction under a tautology condition) prefix is defined as "$<C_{422}, k_{422}, C_{423}>$", and the r_{423} (channel-based calling action under a certain (tautology excluded) condition) prefix is defined as "**if** $cond_{423}$ **then** $<C_{424}, CG, k_{423}>$".

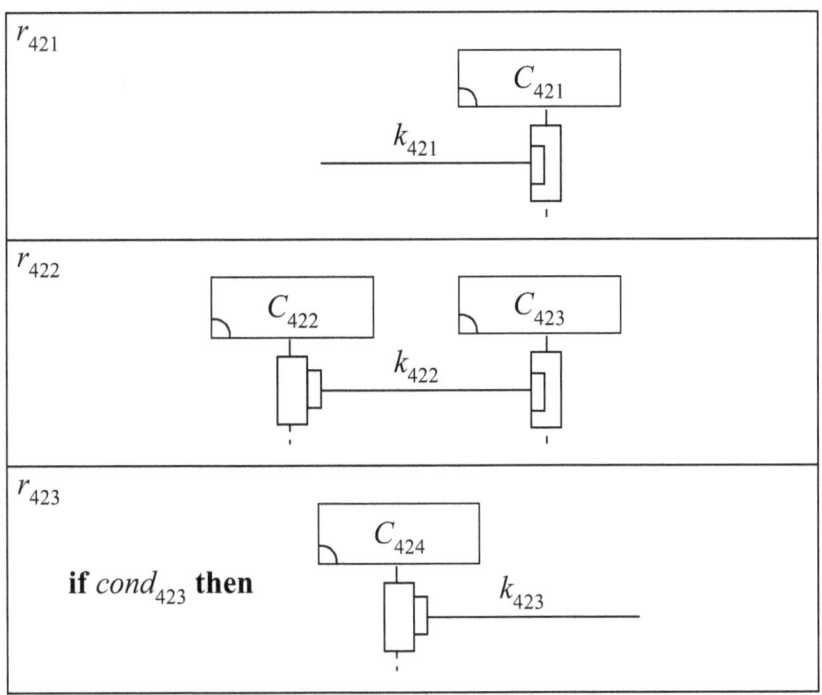

Structural composition of "C_{422}" and "C_{423}" into "C_{425}" means to rename both the "C_{422}" and "C_{423}" component to the "C_{425}" component.

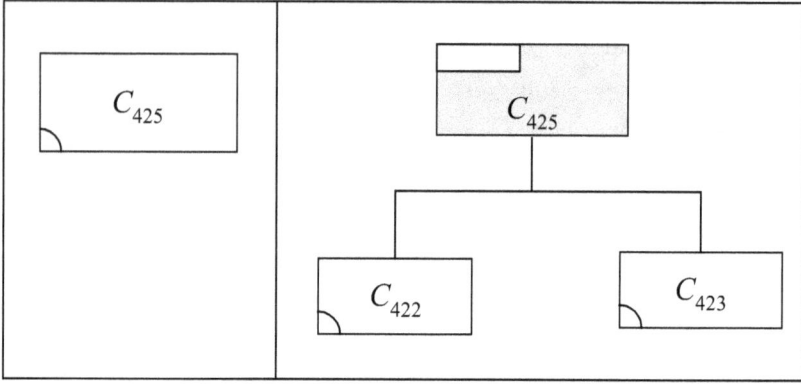

After structurally composing of "C_{422}" and "C_{423}" into "C_{425}", the r_{422} prefix becomes the r_{424} (channel-based interaction under a tautology condition) prefix represented by $<C_{425}, k_{422}, C_{425}>$ which stands for an internal interaction (i.e. λ) because it describes the "C_{425}" component interacts with itself.

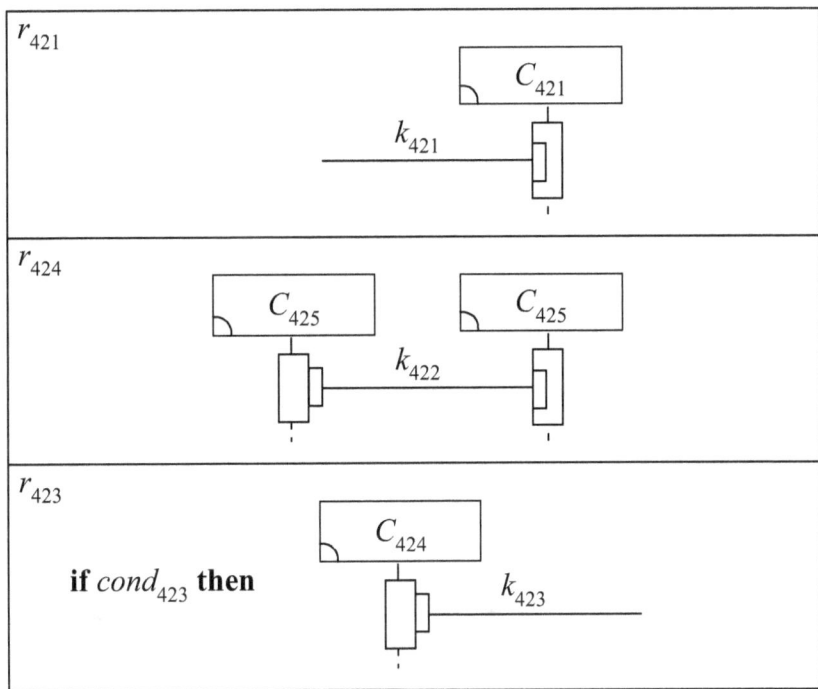

The following transition graph shows the semantics of process A_{421}.

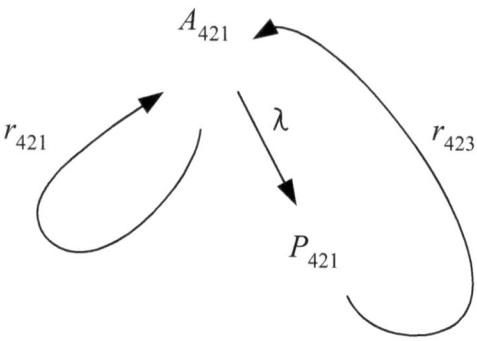

In the transition graph of the A_{421}'s generalized SBC process, processes A_{421} and P_{421} are defined as:

$$A_{421} \overset{\text{def}}{=\joinrel=} r_{421} \bullet A_{421} + \lambda \bullet P_{421}$$

$$P_{421} \overset{\text{def}}{=\joinrel=} r_{423} \bullet A_{421}$$

Operation-Based Structural Composition

As a first example, consider the generalized SBC process Constant A_{431} is defined as "$r_{431} \bullet r_{432} \bullet STOP[C_{434}/C_{432}, C_{434}/C_{433}]$", the r_{431} (operation-based called action under a certain (tautology excluded) condition) prefix is defined as "if $cond_{431}$ **then** <O_C, C_{431}, CD, l_{431}>", and the r_{432} (operation-based interaction under a tautology condition) prefix is defined as "<O_R, C_{432}, l_{432}, C_{433}>".

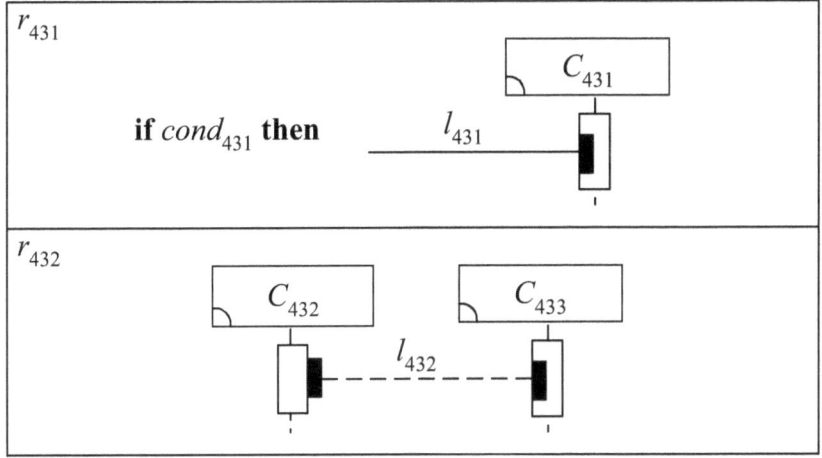

Structural composition of "C_{432}" and "C_{433}" into "C_{434}" means to rename both the "C_{432}" and "C_{433}" component to the "C_{434}" component.

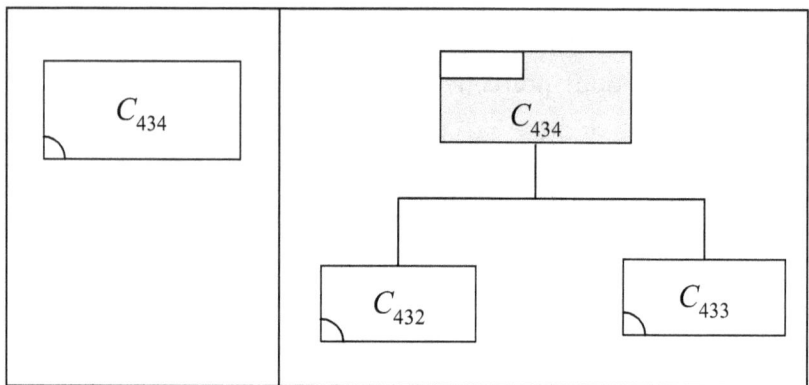

After structurally composing of "C_{432}" and "C_{433}" into "C_{434}", the r_{432} prefix becomes the r_{433} (operation-based interaction under a tautology condition) prefix represented by <O_R, C_{434}, l_{432}, C_{434}> which stands for an internal interaction (i.e. λ) because it describes the "C_{434}" component interacts with itself.

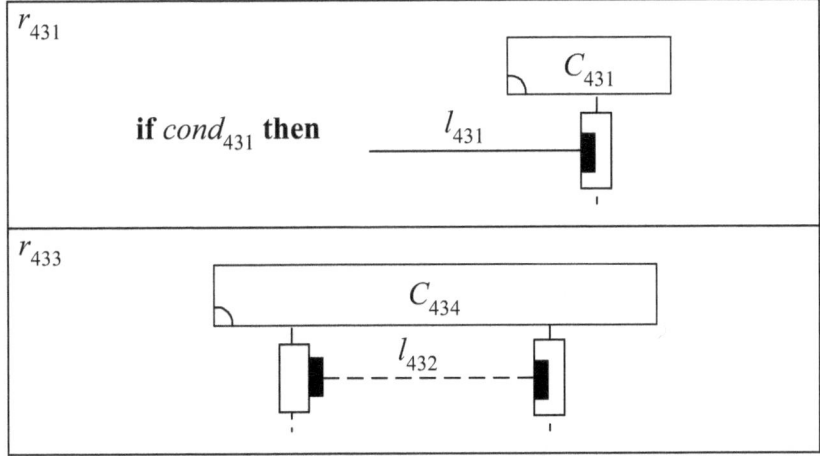

The following transition graph shows the semantics of process A_{431}.

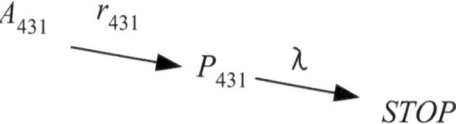

In the transition graph of the A_{431}'s generalized SBC process, processes A_{431} and P_{431} are defined as:

$$A_{431} \stackrel{\text{def}}{=\joinrel=} r_{431} \bullet P_{431}$$

$$P_{431} \stackrel{\text{def}}{=\joinrel=} \lambda \bullet STOP$$

As a second example, consider the generalized SBC process Constant A_{441} is defined as "**fix**$(X_{441}=r_{441} \bullet X_{441}+r_{442} \bullet r_{443} \bullet X_{441})[C_{445}/C_{443}, C_{445}/C_{444}]$" and the r_{441} (operation-based called action under a tautology condition) prefix is defined as $<$O_R, C_{441}, CD, $l_{441}>$, the r_{442} (operation-based calling action under a tautology condition) prefix is defined as $<$O_R, C_{442}, CG, $l_{442}>$, and the r_{443} (operation-based interaction under a certain (tautology excluded) condition) prefix is defined as "**if** $cond_{443}$ **then** $<$O_R, C_{443}, l_{443}, $C_{444}>$".

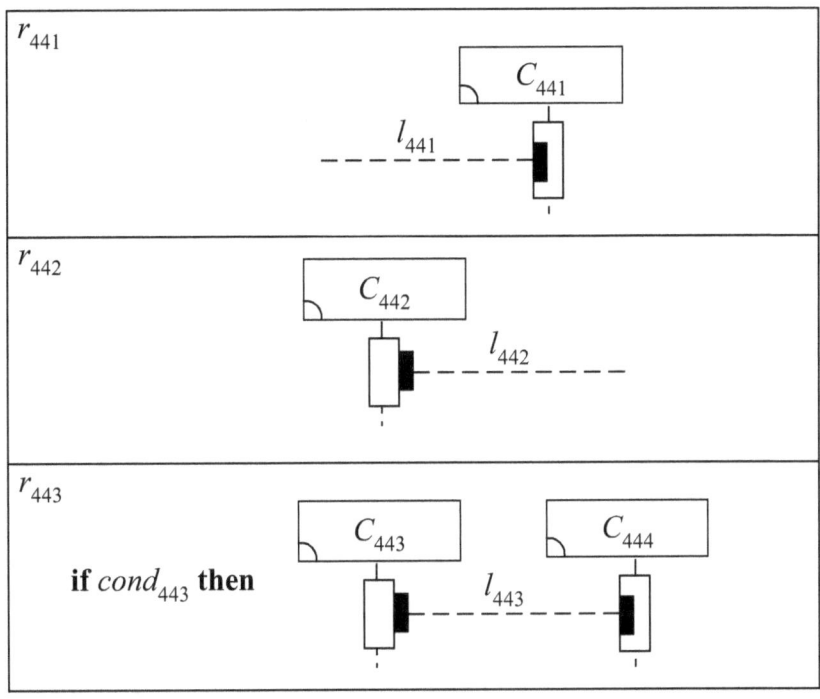

Structural composition of "C_{443}" and "C_{444}" into "C_{445}" means to rename both the "C_{443}" and "C_{444}" component to the "C_{445}" component.

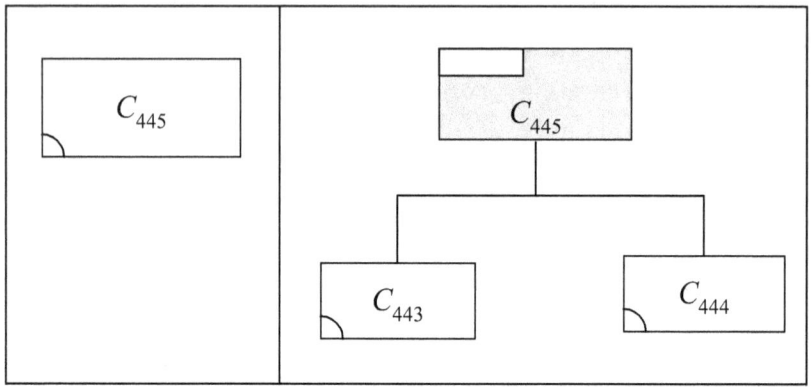

After structurally composing of "C_{443}" and "C_{444}" into "C_{445}", the r_{443} prefix becomes the r_{444} (operation-based interaction under a certain (tautology excluded) condition) prefix represented by "**if** $cond_{443}$ **then** <O_R, C_{445}, l_{443}, C_{445}>" which stands for an internal interaction (i.e. λ) under a certain (tautology excluded) condition. (Be noted that an internal interaction under a certain (tautology excluded) condition is not equal to an internal interaction.)

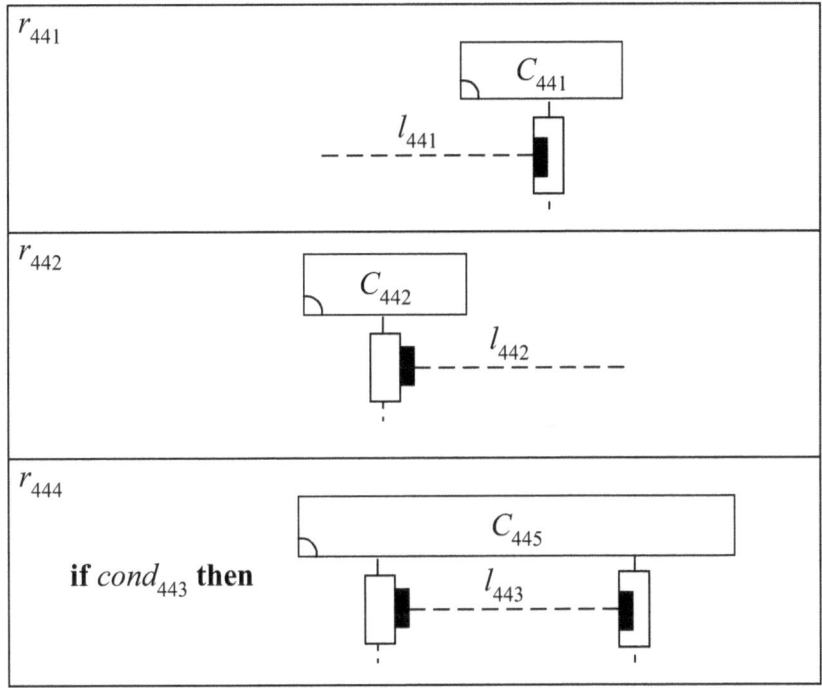

The following transition graph shows the semantics of process A_{441}.

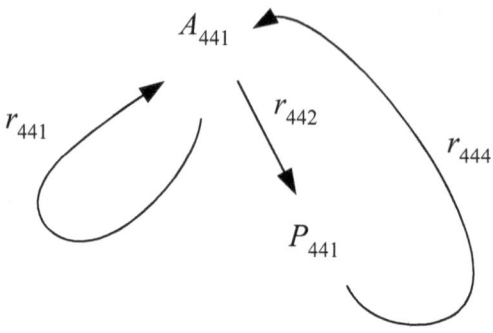

In the transition graph of the A_{441}'s generalized SBC process, processes A_{441} and P_{441} are defined as:

$$A_{441} \stackrel{\text{def}}{=\!=} r_{441} \bullet A_{441} + r_{442} \bullet P_{441}$$

$$P_{441} \stackrel{\text{def}}{=\!=} r_{444} \bullet A_{441}$$

PART XI: OBSERVATION CONGRUENCE THEORY OF GENERALIZED SBC PROCESS ALGEBRA

Preliminary Definitions

A few preliminary definitions are needed. Definition 11.1 and Definition 11.2 are easy to comprehend.

Definition 11.1 $\quad \hat{r} \; = \; \varepsilon$ (the empty sequence), if $r = \lambda$

$\qquad\qquad\qquad = r$, if $r \neq \lambda$

Definition 11.2 $\quad E \stackrel{r}{\Longrightarrow} E' \quad$ iff $\quad E\,(\stackrel{\lambda}{\rightarrow})^* \stackrel{r}{\rightarrow} (\stackrel{\lambda}{\rightarrow})^* E'$

$\qquad\qquad\quad E \stackrel{\varepsilon}{\Longrightarrow} E' \quad$ iff $\quad E\,(\stackrel{\lambda}{\rightarrow})^* (\stackrel{\lambda}{\rightarrow})^* E'$

Observation Equivalence

To achieve the observation congruence [Chao15c, Chao15g, Chao15h, Miln89, Miln99], we need to define the observation equivalence first. Definition 11.3 and Definition 11.4 together define the observation equivalence.

Definition 11.3   A binary relation $S \subseteq \Pi \times \Pi$ over generalized SBC processes is a *bisimulation* if $(P, Q) \in S$ implies, for all or each $r \in R$,

(i) whenever $P \xrightarrow{r} P'$ then, for some Q', $Q \overset{\widehat{r}}{\Longrightarrow} Q'$ and $(P', Q') \in S$

(ii) whenever $Q \xrightarrow{r} Q'$ then, for some P', $P \overset{\widehat{r}}{\Longrightarrow} P'$ and $(P', Q') \in S$

Definition 11.4 P and Q are observation equivalent, written $P \approx Q$, if $(P, Q) \in S$ for some bisimulation S. That is,

$$\approx = \bigcup (S : S \text{ is a bisimulation})$$

Observation Congruence

Once we have the definition of observation equivalence, i.e. $\approx$, we shall use it to define the observation congruence i.e. =.

Definition 11.5 P and Q are observation congruent, written $P = Q$, if for all r

(i) whenever $P \xrightarrow{r} P'$ then, for some Q', $Q \Longrightarrow Q'$ and $P' \approx Q'$;

(ii) whenever $Q \xrightarrow{r} Q'$ then, for some P', $P \Longrightarrow P'$ and $P' \approx Q'$.

PART XII: EXAMPLES OF OBSERVATION CONGRUENCE

As a first example, consider the generalized SBC process P_{501} is defined as "$(r_{501} \bullet r_{502} \bullet STOP)$" and the r_{501} (channel-based called action under a certain (tautology excluded) condition) prefix is defined as "if $cond_{501}$ **then** $<C_{501}, CD, k_{501}>$" and the r_{502} (channel-based internal interaction under a tautology condition) prefix is defined as "$<C_{502}, k_{502}, C_{502}>$".

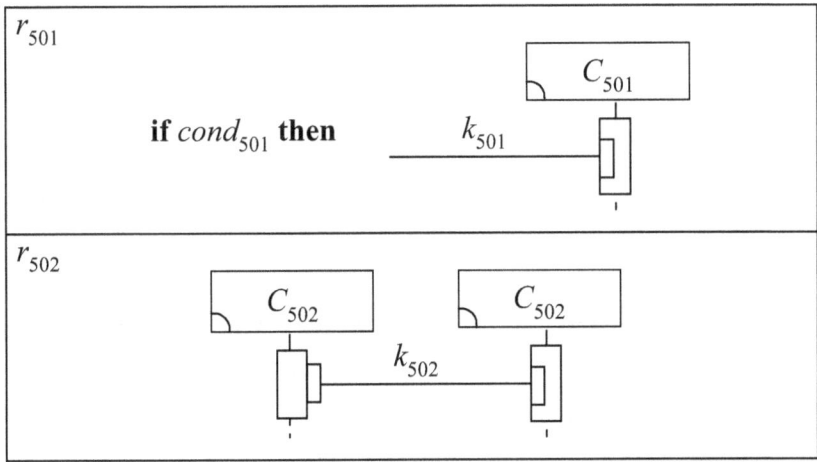

The following transition graph shows the semantics of process P_{501}.

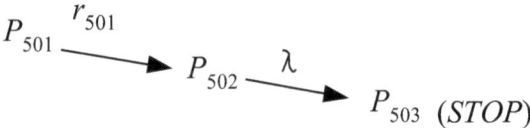

In the transition graph of the P_{501}'s generalized SBC process, processes P_{501}, P_{502} and P_{503} are defined as:

$$P_{501} \overset{\text{def}}{=\joinrel=} r_{501} \bullet P_{502}$$

$$P_{502} \overset{\text{def}}{=\joinrel=} \lambda \bullet P_{503}$$

$$P_{503} \overset{\text{def}}{=\joinrel=} STOP$$

Also consider the generalized SBC process Q_{501} is defined as "$(r_{501} \bullet STOP)$" the r_{501} (channel-based called action under a certain (tautology excluded) condition) prefix is defined as "**if** $cond_{501}$ **then** $<C_{501}, CD, k_{501}>$".

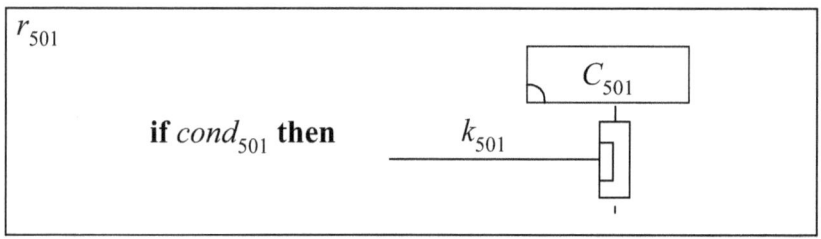

The following transition graph shows the semantics of process Q_{501}.

In the transition graph of the Q_{501}'s generalized SBC process, processes Q_{501} and Q_{502} are defined as:

$$Q_{501} \stackrel{\text{def}}{=\!=} r_{501} \bullet Q_{502}$$

$$Q_{502} \stackrel{\text{def}}{=\!=} STOP$$

We can easily verify that $S_{501} = \{(P_{501}, Q_{501}), (P_{502}, Q_{502}), (P_{503}, Q_{502})\}$ is a bisimulation.

Using the S_{501} bisimulation, we then are able to verify that P_{501} and Q_{501} are observation congruent because (1) $P_{501} \xrightarrow{r_{501}} P_{502}$, then we have Q_{502} that $Q_{501} \xRightarrow{r_{501}} Q_{502}$ and $P_{502} \approx Q_{502}$, and (2) $Q_{501} \xrightarrow{r_{501}} Q_{502}$, then we have P_{502} that $P_{501} \xRightarrow{r_{501}} P_{502}$ and $P_{502} \approx Q_{502}$.

As a second example, consider the generalized SBC process P_{511} is defined as "$(r_{511} \bullet r_{512} \bullet STOP)$" and the r_{511} (operation-based called action under a tautology condition) prefix is defined as "$<O\_C, C_{511}, CD, l_{511}>$" and the r_{512} (operation-based internal interaction under a certain (tautology excluded) condition) prefix is defined as "**if** $cond_{512}$ **then** $<O\_C, C_{512}, l_{512}, C_{512}>$". (Be noted that an internal interaction under a certain (tautology excluded) condition is not equal to an internal interaction.)

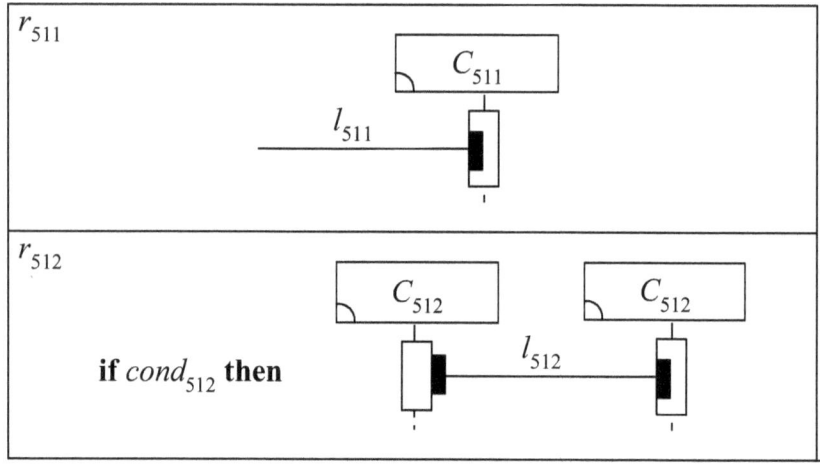

The following transition graph shows the semantics of process P_{511}.

$$P_{511} \xrightarrow{\ r_{511}\ } P_{512} \xrightarrow{\ r_{512}\ } P_{513}\ (STOP)$$

In the transition graph of the P_{511}'s generalized SBC process, processes P_{511}, P_{512} and P_{513} are defined as:

$$P_{511} \overset{\text{def}}{=\joinrel=} r_{511} \bullet P_{512}$$

$$P_{512} \overset{\text{def}}{=\joinrel=} r_{512} \bullet P_{513}$$

$$P_{513} \overset{\text{def}}{=\joinrel=} STOP$$

Also consider the generalized SBC process Q_{511} is defined as "$(r_{511} \bullet STOP)$" and the r_{511} (operation-based called action under a tautology condition) prefix is defined as "<O_C, C_{511}, CD, l_{511}>".

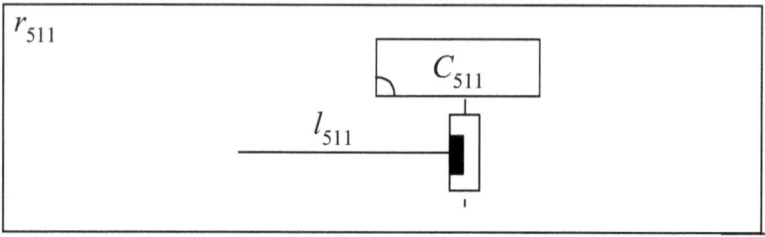

The following transition graph shows the semantics of process Q_{511}.

$$Q_{511} \xrightarrow{r_{511}} Q_{512} \ (STOP)$$

In the transition graph of the Q_{511}'s generalized SBC process, processes Q_{511} and Q_{512} are defined as:

$$Q_{511} \stackrel{\text{def}}{=\joinrel=} r_{511} \bullet Q_{512}$$

$$Q_{512} \stackrel{\text{def}}{=\joinrel=} STOP$$

We can easily verify that $S_{511} = \{(P_{513}, Q_{512})\}$ is a bisimulation.

Using the S_{511} bisimulation, we then are able to verify that P_{511} and Q_{511} are not observation congruent because for P_{511} $\xrightarrow{r_{511}} P_{512}$ and we find no Q_{kkk} that $Q_{511} \overset{r_{511}}{\Longrightarrow} Q_{kkk}$ and $P_{512} \overset{\sim}{\sim} Q_{kkk}$.

As a third example, consider the generalized SBC process P_{515} is defined as "$(r_{515} \bullet r_{516} \bullet STOP)$" and the r_{515} (operation-based called action under a tautology condition) prefix is defined as "$<O\_C, C_{515}, CD, l_{515}>$" and the r_{516} (operation-based internal interaction under a certain (tautology excluded) condition) prefix is defined as "**if** $cond_{516}$ **then** $<O\_C, C_{516}, l_{516}, C_{516}>$". (Be noted that an internal interaction under a certain (tautology excluded) condition is not equal to an internal interaction.)

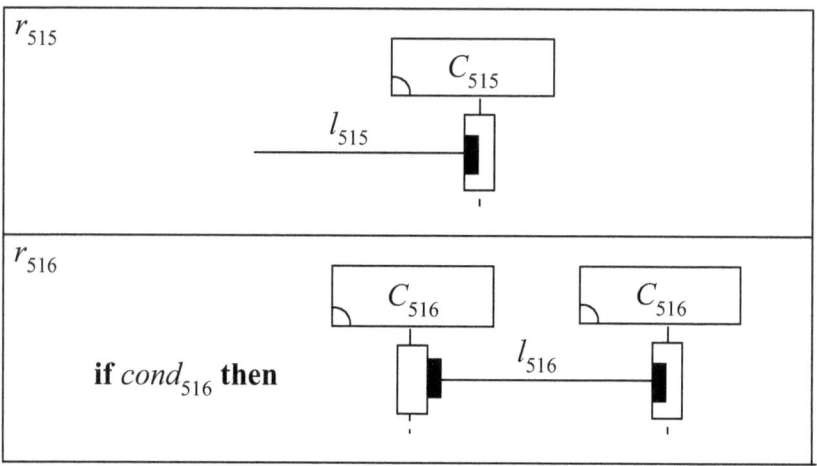

The following transition graph shows the semantics of process P_{515}.

$$P_{515} \xrightarrow{r_{515}} P_{518} \xrightarrow{r_{516}} P_{519} \ (STOP)$$

In the transition graph of the P_{515}'s generalized SBC process, processes P_{515}, P_{518} and P_{519} are defined as:

$$P_{515} \stackrel{\text{def}}{=\!=} r_{515} \bullet P_{518}$$

$$P_{518} \stackrel{\text{def}}{=\!=} r_{516} \bullet P_{519}$$

$$P_{519} \stackrel{\text{def}}{=\!=} STOP$$

Also consider the generalized SBC process Q_{515} is defined as "$(r_{515} \bullet r_{517} \bullet STOP)$" and the r_{515} (operation-based called action under a tautology condition) prefix is defined as "<O_C, C_{515}, CD, l_{515}>" and the r_{517} (operation-based internal interaction under a certain (tautology excluded) condition) prefix is defined as "**if** $cond_{516}$ **then** <O_C, C_{517}, l_{517}, C_{517}>". (Be noted that an internal interaction under a certain (tautology excluded) condition is not equal to an internal interaction.)

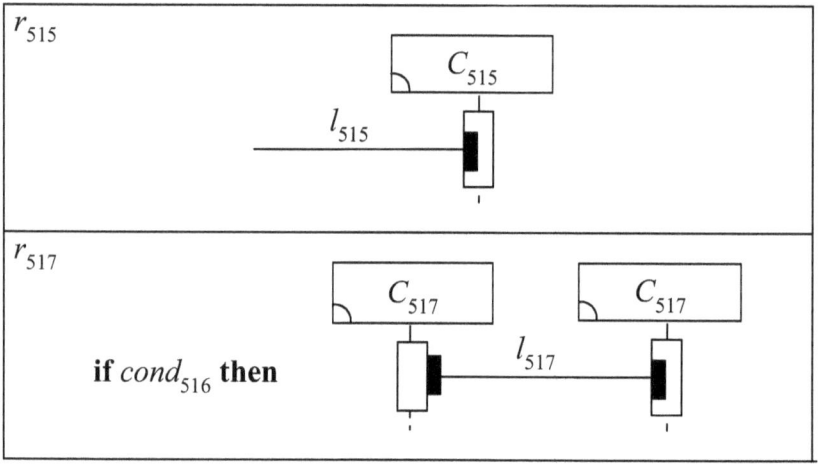

The following transition graph shows the semantics of process Q_{515}.

$$Q_{515} \xrightarrow{r_{515}} Q_{518} \xrightarrow{r_{517}} Q_{519} \; (STOP)$$

In the transition graph of the Q_{515}'s generalized SBC process, processes Q_{515}, Q_{518} and Q_{519} are defined as:

$$Q_{515} \overset{\text{def}}{=\joinrel=} r_{515} \bullet Q_{518}$$

$$Q_{518} \overset{\text{def}}{=\joinrel=} r_{517} \bullet Q_{519}$$

$$Q_{519} \overset{\text{def}}{=\joinrel=} STOP$$

We can easily verify that $S_{515} = \{(P_{515}, Q_{515}), (P_{518}, Q_{518}), (P_{519}, Q_{519})\}$ is a bisimulation.

Using the S_{515} bisimulation, we then are able to verify that P_{515} and Q_{515} are observation congruent because (1) $P_{515} \xrightarrow{r_{515}} P_{5158}$, then we have Q_{518} that $Q_{515} \overset{r_{515}}{\Longrightarrow} Q_{518}$ and $P_{518} \approx Q_{518}$, and (2) $Q_{515} \xrightarrow{r_{515}} Q_{518}$, then we have P_{518} that $P_{515} \overset{r_{515}}{\Longrightarrow} P_{518}$ and $P_{518} \approx Q_{518}$.

As a fourth example, consider the generalized SBC process P_{521} is defined as "$(r_{521} \bullet r_{522} \bullet STOP)$" and the r_{521} (channel-based internal interaction under a tautology condition) prefix is defined as "$<C_{521}, k_{521}, C_{521}>$" and the r_{522} (channel-based called action under a certain (tautology excluded) condition) prefix is defined as "**if** $cond_{522}$ **then** $<C_{522}, CD, k_{522}>$".

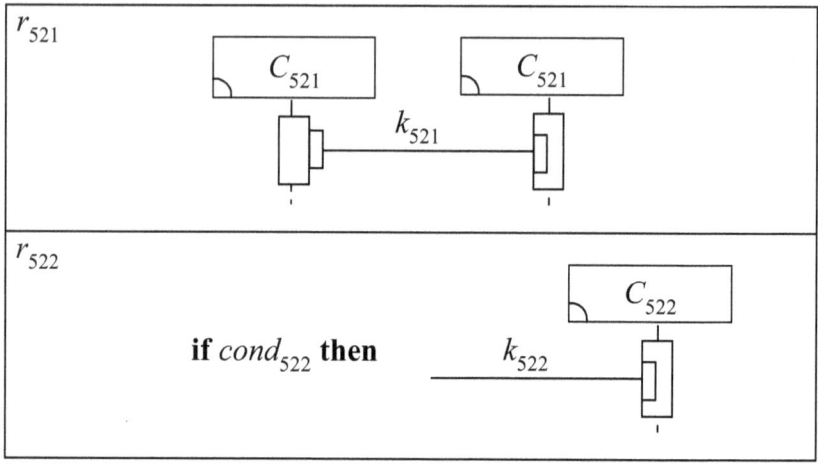

The following transition graph shows the semantics of process P_{521}.

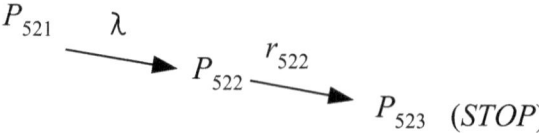

In the transition graph of the P_{521}'s generalized SBC process, processes P_{521}, P_{522} and P_{523} are defined as:

$$P_{521} \stackrel{\text{def}}{=\joinrel=} \lambda \bullet P_{522}$$

$$P_{522} \stackrel{\text{def}}{=\joinrel=} r_{522} \bullet P_{523}$$

$$P_{523} \stackrel{\text{def}}{=\joinrel=} STOP$$

Also consider the generalized SBC process Q_{521} is defined as "$(r_{522} \bullet STOP)$" and the r_{522} (channel-based called action under a certain (tautology excluded) condition) prefix is defined as "**if** $cond_{522}$ **then** $<C_{522}, CD, k_{522}>$".

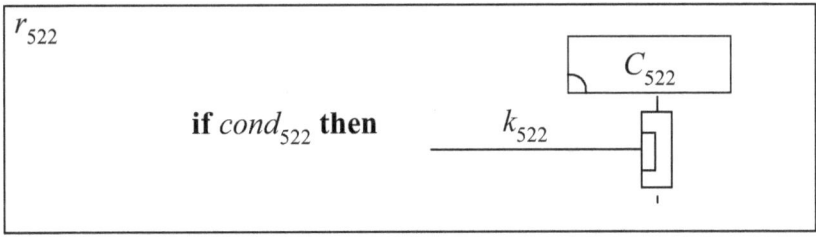

The following transition graph shows the semantics of process Q_{521}.

$$Q_{521} \xrightarrow{\quad r_{522} \quad} Q_{522} \ (STOP)$$

In the transition graph of the Q_{521}'s generalized SBC process, processes Q_{521} and Q_{522} are defined as:

$$Q_{521} \overset{def}{=\!=} r_{522} \bullet Q_{522}$$

$$Q_{522} \overset{def}{=\!=} STOP$$

We can easily verify that $S_{521} = \{(P_{521}, Q_{521}), (P_{522}, Q_{521}), (P_{523}, Q_{522})\}$ is a bisimulation.

Using the S_{521} bisimulation, we then are able to verify that P_{521} and Q_{521} are not observation congruent because for $P_{521} \xrightarrow{\lambda} P_{522}$ and we find no Q_{kkk} that $Q_{521} \overset{\lambda}{\Longrightarrow} Q_{kkk}$ and $P_{522} \overset{\sim}{\approx} Q_{kkk}$.

As a fifth example, consider the generalized SBC process P_{531} is defined as "$\mathbf{fix}(X_{531}=r_{531} \bullet X_{531}+r_{532} \bullet r_{533} \bullet X_{531})$" and the r_{531} (channel-based called action under a tautology condition) prefix is defined as " $<C_{531}$, CD, $k_{531}>$" and the r_{532} (channel-based internal interaction under a tautology condition) prefix is defined as "$<C_{532}$, k_{532}, $C_{532}>$" and the r_{533} (channel-based called action under a tautology condition) prefix is defined as " $<C_{533}$, CD, $k_{533}>$".

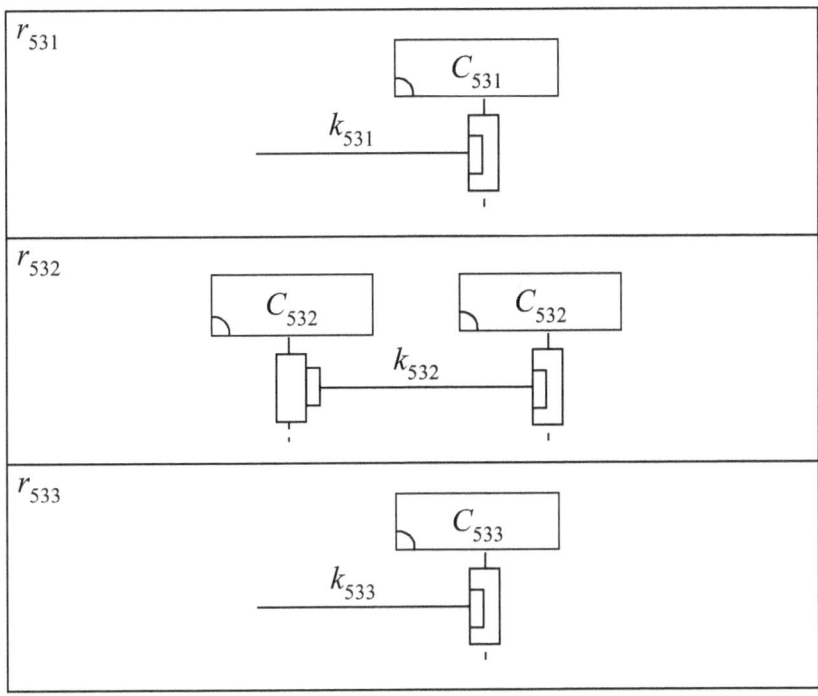

The following transition graph shows the semantics of process P_{531}.

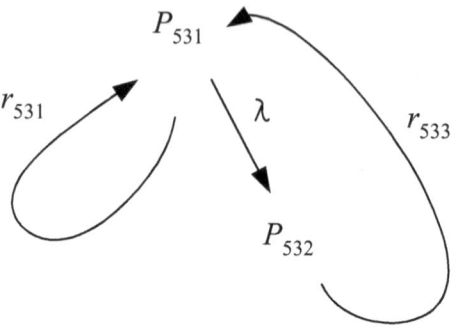

In the transition graph of the P_{531}'s generalized SBC process, processes P_{531} and P_{532} are defined as:

$$P_{531} \overset{\text{def}}{=\!=} r_{531} \bullet P_{531} + \lambda \bullet P_{532}$$

$$P_{532} \overset{\text{def}}{=\!=} r_{533} \bullet P_{531}$$

Also consider the generalized SBC process Q_{531} is defined as "**fix**$(X_{534}=r_{531} \bullet X_{531}+r_{533} \bullet X_{534})$" and the r_{531} (channel-based called action under a tautology condition) prefix is defined as " $<C_{531}$, CD, $k_{531}>$" and the r_{533} (channel-based called action under a tautology condition) prefix is defined as " $<C_{533}$, CD, $k_{533}>$".

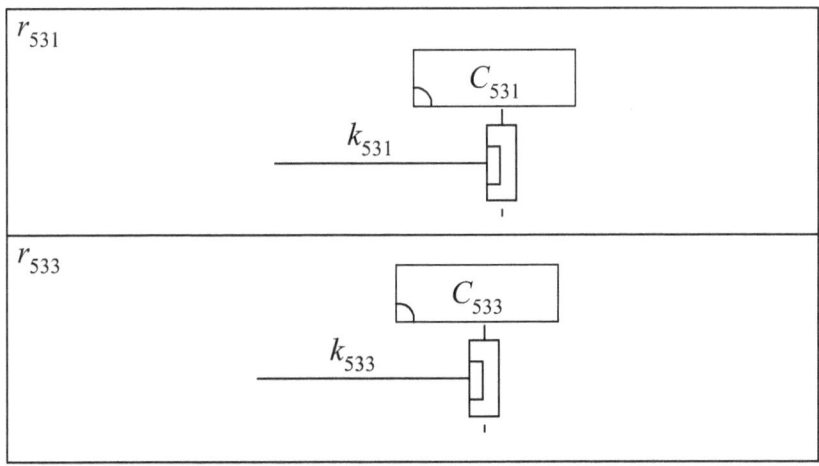

The following transition graph shows the semantics of process Q_{531}.

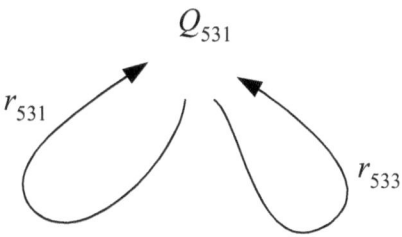

In the transition graph of the Q_{531}'s generalized SBC process, process Q_{531} is defined as:

$$Q_{531} \overset{\text{def}}{=\!=} r_{531} \bullet Q_{531} + r_{533} \bullet Q_{531}$$

We can easily verify that a bisimulation $S_{531} = \{\ \}$ is an empty set.

Using the S_{531} bisimulation, we then are able to verify that P_{531} and Q_{531} are not observation congruent because for P_{531} $\xrightarrow{r_{531}} P_{531}$ and we find no Q_{kkk} that $Q_{531} \overset{r_{531}}{=\!=\!\Longrightarrow} Q_{kkk}$ and $P_{531} \overset{\sim}{\approx} Q_{kkk}$.

As a sixth example, consider the generalized SBC process P_{541} is defined as "$\mathbf{fix}(X_{541}= r_{541}\bullet X_{541}+r_{542}\bullet r_{543}\bullet X_{541})$" and the r_{541} (channel-based called action under a certain (tautology excluded) condition) prefix is defined as "if $cond_{541}$ **then** $<C_{541}$, CD, $k_{541}>$" and the r_{542} (channel-based called action under a certain (tautology excluded) condition) prefix is defined as "if $cond_{542}$ **then** $<C_{542}$, CD, $k_{542}>$" and the r_{543} (channel-based internal interaction under a tautology condition) prefix is defined as "$<C_{543}$, k_{543}, $C_{543}>$".

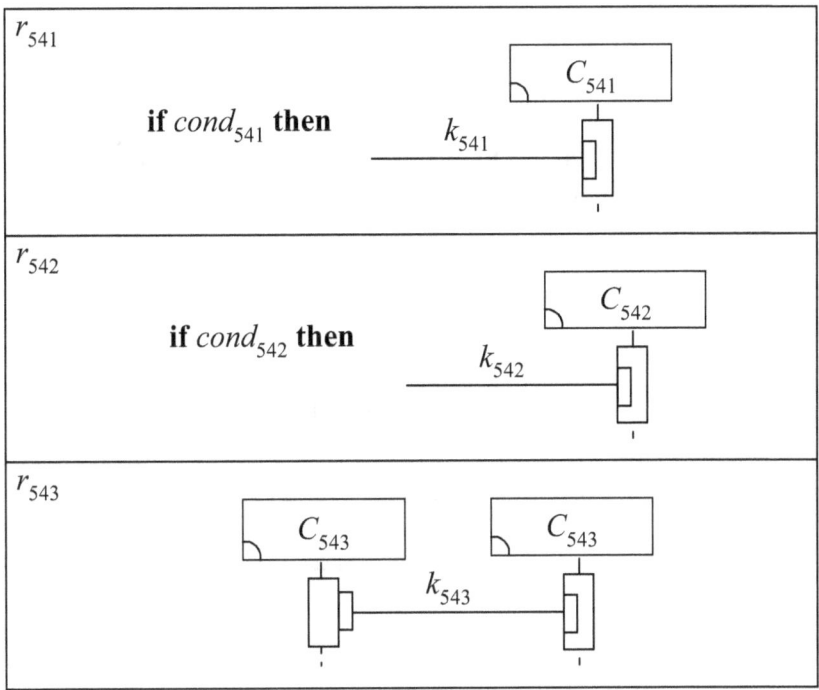

The following transition graph shows the semantics of process P_{541}.

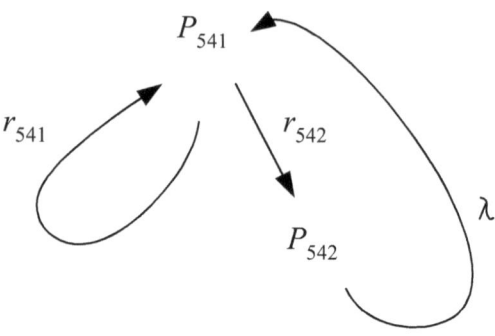

In the transition graph of the P_{541}'s generalized SBC process, processes P_{541} and P_{542} are defined as:

$$P_{541} \overset{\text{def}}{=\!=} r_{541} \bullet P_{541} + r_{542} \bullet P_{542}$$

$$P_{542} \overset{\text{def}}{=\!=} \lambda \bullet P_{541}$$

Also consider the generalized SBC process Q_{541} is defined as "**fix**$(X_{544}=r_{541}\bullet X_{544}+r_{542}\bullet X_{544})$" and the r_{541} (channel-based called action under a certain (tautology excluded) condition) prefix is defined as "**if** $cond_{541}$ **then** $<C_{541}, CD, k_{541}>$" and the r_{542} (channel-based called action under a certain (tautology excluded) condition) prefix is defined as "**if** $cond_{542}$ **then** $<C_{542}, CD, k_{542}>$".

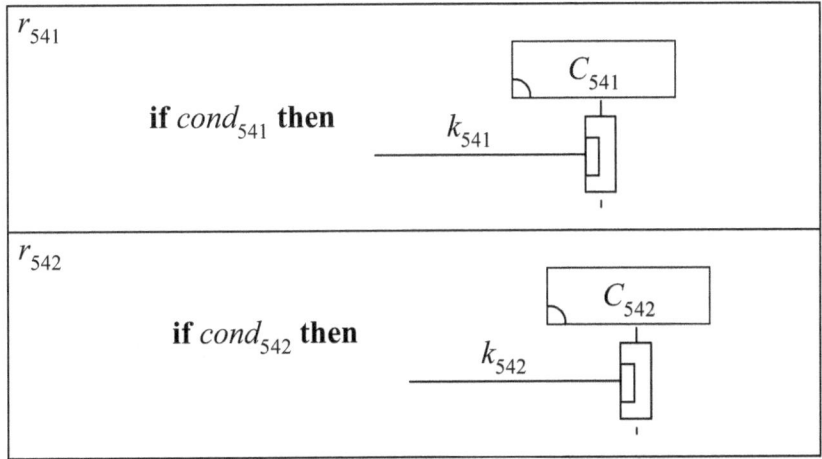

The following transition graph shows the semantics of process Q_{541}.

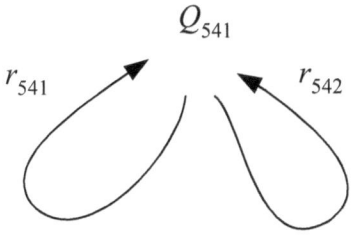

In the transition graph of the Q_{541}'s generalized SBC process, process Q_{541} is defined as:

$$Q_{541} \stackrel{\text{def}}{=\!=} r_{541} \bullet Q_{541} + r_{542} \bullet Q_{541}$$

We can easily verify that $S_{541} = \{(P_{541}, Q_{541}), (P_{542}, Q_{541})\}$ is a bisimulation.

Using the S_{541} bisimulation, we then are able to verify that P_{541} and Q_{541} are observation congruent because (1) $P_{541} \xrightarrow{r_{541}} P_{541}$, then we have Q_{541} that $Q_{541} \overset{r_{541}}{=\!=\!\Longrightarrow} Q_{541}$ and $P_{541} \approx Q_{541}$, and (2) $P_{541} \xrightarrow{r_{542}} P_{542}$, then we have Q_{541} that $Q_{541} \overset{r_{542}}{=\!=\!\Longrightarrow} Q_{541}$ and $P_{542} \approx Q_{541}$, and (3) $Q_{541} \xrightarrow{r_{541}} Q_{541}$, then we have P_{541} that $P_{541} \overset{r_{541}}{=\!=\!\Longrightarrow} P_{541}$ and $P_{541} \approx Q_{541}$, and (4) $Q_{541} \xrightarrow{r_{542}} Q_{541}$, then we have P_{542} that $P_{541} \overset{r_{542}}{=\!=\!\Longrightarrow} P_{542}$ and $P_{542} \approx Q_{541}$.

As a seventh example, consider the generalized SBC process P_{551} is defined as "$r_{551} \bullet r_{552} \bullet \mathbf{fix}(X_{551} = r_{551} \bullet r_{553} \bullet r_{552} \bullet X_{551} + r_{553} \bullet r_{551} \bullet r_{552} \bullet X_{551})$" and the r_{551} (channel-based called action under a certain (tautology excluded) condition) prefix is defined as "if $cond_{551}$ then $<C_{551}$, CD, $k_{551}>$" and the r_{552} (channel-based internal interaction under a tautology condition) prefix is defined as "$<C_{552}$, k_{552}, $C_{552}>$" and the r_{553} (channel-based calling action under a certain (tautology excluded) condition) prefix is defined as "if $cond_{553}$ then $<C_{553}$, CG, $k_{553}>$".

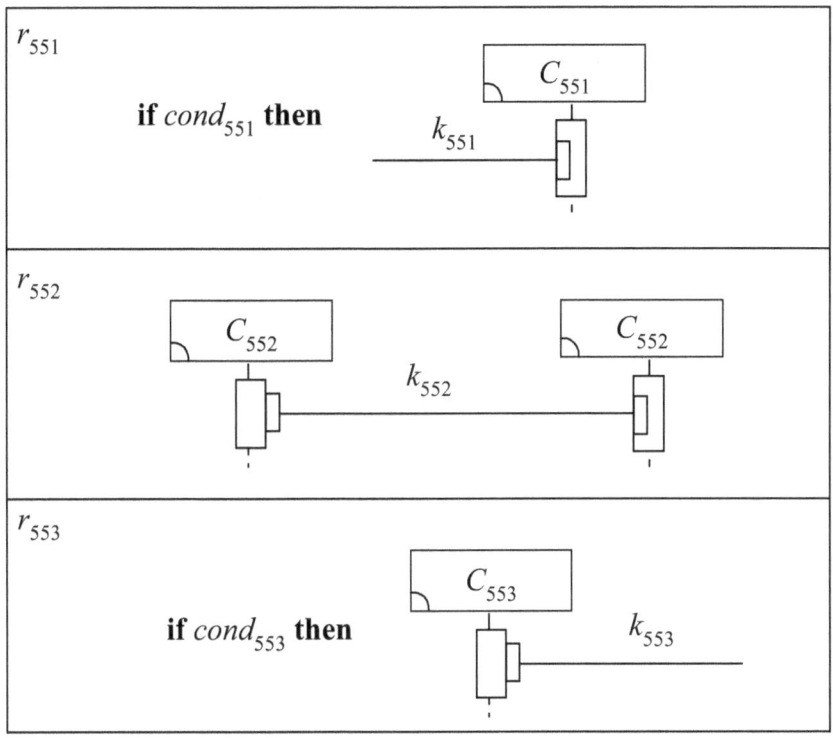

The following transition graph shows the semantics of process P_{551}.

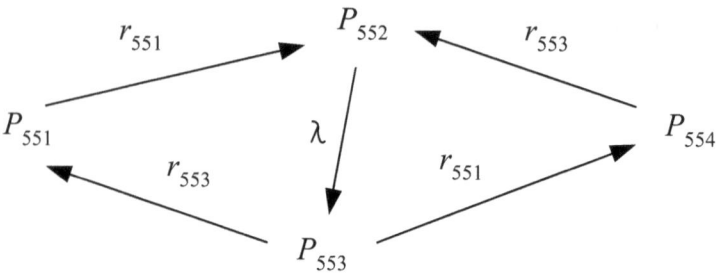

In the transition graph of the P_{551}'s generalized SBC process, processes P_{551}, P_{552}, P_{553} and P_{554} are defined as:

$$P_{551} \stackrel{\text{def}}{=\joinrel=} r_{551} \bullet P_{552}$$

$$P_{552} \stackrel{\text{def}}{=\joinrel=} \lambda \bullet P_{553}$$

$$P_{553} \stackrel{\text{def}}{=\joinrel=} r_{551} \bullet P_{554} + r_{553} \bullet P_{551}$$

$$P_{554} \stackrel{\text{def}}{=\joinrel=} r_{553} \bullet P_{552}$$

Also consider the generalized SBC process Q_{551} is defined as "$r_{551} \bullet \mathbf{fix}(X_{552} = r_{551} \bullet r_{553} \bullet X_{552} + r_{553} \bullet r_{551} \bullet X_{552})$" and the r_{551} (channel-based called action under a certain (tautology excluded) condition) prefix is defined as "$\mathbf{if}\ cond_{551}\ \mathbf{then}\ <C_{551}, \mathrm{CD}, k_{551}>$" and the r_{553} (channel-based calling action under a certain (tautology excluded) condition) prefix is defined as "$\mathbf{if}\ cond_{553}\ \mathbf{then}\ <C_{553}, \mathrm{CG}, k_{553}>$".

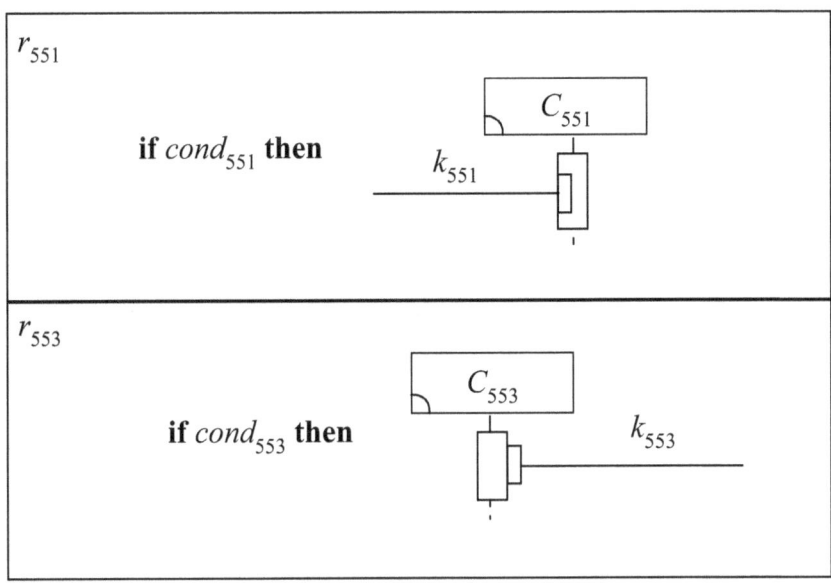

The following transition graph shows the semantics of process Q_{551}.

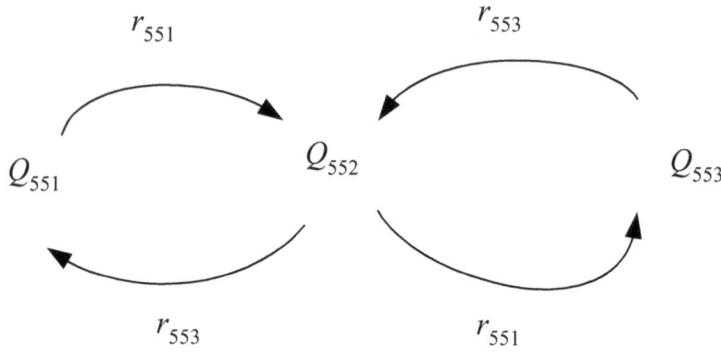

In the transition graph of the Q_{551}'s generalized SBC process, processes Q_{551}, Q_{552} and Q_{553} are defined as:

$$Q_{551} \stackrel{\text{def}}{=\joinrel=} r_{551} \bullet Q_{552}$$

$$Q_{552} \stackrel{\text{def}}{=\joinrel=} r_{551} \bullet Q_{553} + r_{553} \bullet Q_{551}$$

$$Q_{553} \stackrel{\text{def}}{=\joinrel=} r_{553} \bullet Q_{552}$$

We can easily verify that $S_{551} = \{(P_{551}, Q_{551}), (P_{552}, Q_{552}),$ $(P_{553}, Q_{552}), (P_{554}, Q_{553})\}$ is a bisimulation.

Using the S_{551} bisimulation, we then are able to verify that P_{551} and Q_{551} are observation congruent because (1) $P_{551} \xrightarrow{r_{551}} P_{552}$, then we have Q_{552} that $Q_{551} \overset{r_{551}}{\Longrightarrow} Q_{552}$ and P_{552} $\approx Q_{552}$, and (2) $Q_{551} \xrightarrow{r_{551}} Q_{552}$, then we have P_{552} that P_{551} $\overset{r_{551}}{\Longrightarrow} P_{552}$ and $P_{552} \approx Q_{552}$.

PART XIII: LIMITATIONS OF OBSERVATION CONGRUENCE THEORY

Limitation on Observation Equivalence Verification

The observation congruence theory in this book has some limitations. That is, two observation equivalent processes may be verified to be not observation equivalent. We will demonstrate a few examples here. We just leave this limitation in this book until we find the right solution.

As a first example, consider the generalized SBC process P_{601} is defined as "$(r_{601} \bullet r_{602} \bullet STOP)$" and the r_{601} (channel-based internal interaction under a tautology condition) prefix is defined as "$<C_{601}, k_{601}, C_{601}>$" and the r_{602} (channel-based called action under a tautology condition) prefix is defined as "$<C_{602}, CD, k_{602}>$".

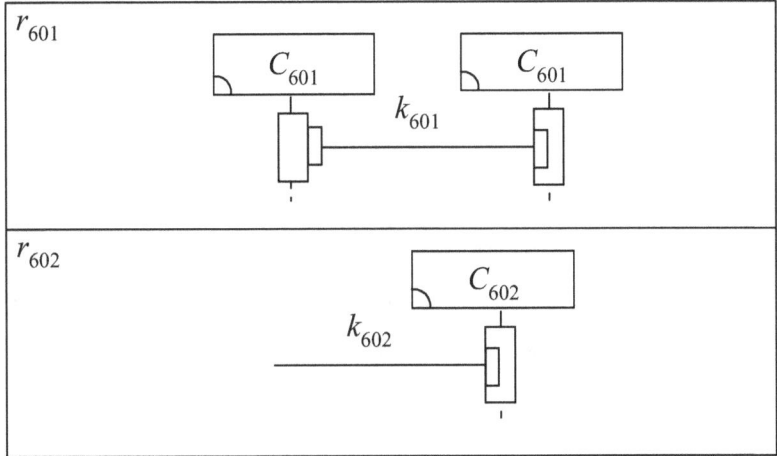

The following transition graph shows the semantics of process P_{601}.

$$P_{601} \quad \lambda$$
$$\longrightarrow P_{602} \xrightarrow{\; r_{602} \;} P_{603} \quad (STOP)$$

In the transition graph of the P_{601}'s generalized SBC process, processes P_{601}, P_{602} and P_{603} are defined as:

$$P_{601} \stackrel{\text{def}}{=\!=} \lambda \bullet P_{602}$$

$$P_{602} \stackrel{\text{def}}{=\!=} r_{602} \bullet P_{603}$$

$$P_{603} \stackrel{\text{def}}{=\!=} STOP$$

Also consider the generalized SBC process Q_{601} is defined as "$(r_{603} \bullet STOP + r_{604} \bullet STOP)$" and the r_{603} (channel-based called action under a certain (tautology excluded) condition) prefix is defined as "**if** $(x < 100)$ **then** $<C_{602}, CD, k_{602}>$" and the r_{604} (channel-based called action under a certain (tautology excluded) condition) prefix is defined as "**if** $(100 <= x)$ **then** $<C_{602}, CD, k_{602}>$".

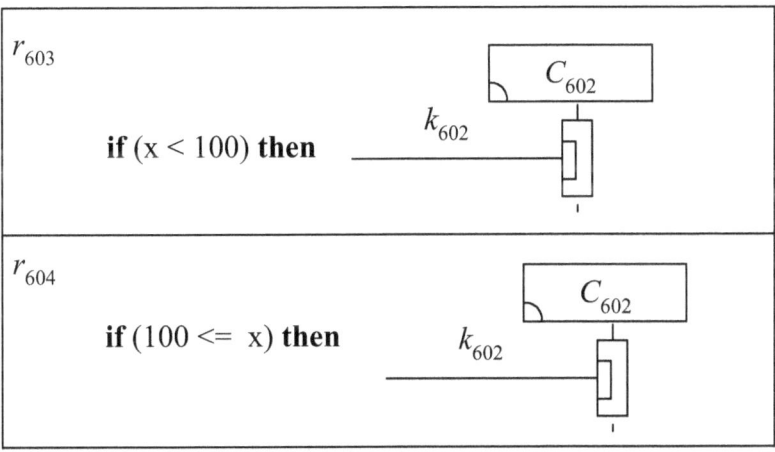

The following transition graph shows the semantics of process Q_{601}.

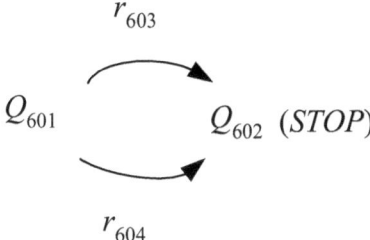

In the transition graph of the Q_{601}'s generalized SBC process, processes Q_{601} and Q_{602} are defined as:

$$Q_{601} \overset{\text{def}}{=} r_{603} \bullet Q_{602} + r_{604} \bullet Q_{602}$$

$$Q_{602} \overset{\text{def}}{=} STOP$$

Although P_{601} and Q_{601} are observation equivalent, in this book we can only verify that they are not observation equivalent.

As a second example, consider the generalized SBC process P_{611} is defined as "$(r_{611} \bullet STOP)$" and the r_{611} (channel-based called action under a tautology condition) prefix is defined as "$<C_{611}$, CD, $k_{611}>$".

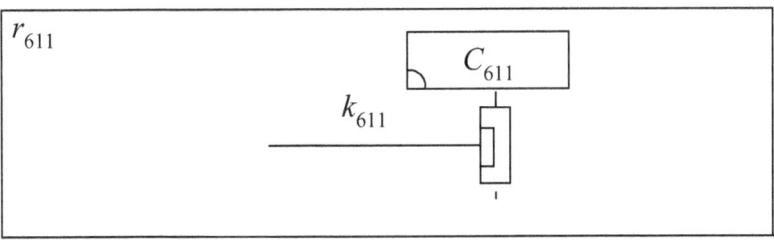

The following transition graph shows the semantics of process P_{611}.

$$P_{611} \xrightarrow{r_{611}} P_{612} \ (STOP)$$

In the transition graph of the P_{601}'s generalized SBC process, processes P_{611} and P_{612} are defined as:

$$P_{611} \stackrel{\text{def}}{=\!=} r_{611} \bullet P_{612}$$

$$P_{612} \stackrel{\text{def}}{=\!=} STOP$$

Also consider the generalized SBC process Q_{611} is defined as "$(r_{612}+r_{613}) \bullet r_{611} \bullet STOP$" and the r_{612} (channel-based internal interaction under a certain (tautology excluded) condition) prefix is defined as "**if** (y < 200) **then** $<C_{612}, k_{612}, C_{612}>$" the r_{613} (channel-based internal interaction under a certain (tautology excluded) condition) prefix is defined as " (200 <= y) **then** $<C_{613}, k_{613}, C_{613}>$" and the r_{611} (channel-based called action under a tautology condition) prefix is defined as "$<C_{611}, CD, k_{611}>$".

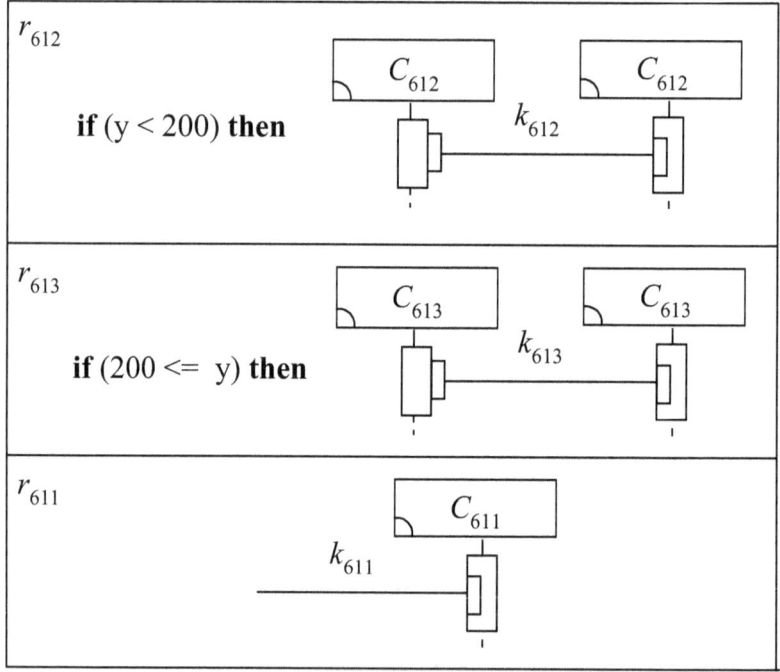

The following transition graph shows the semantics of process Q_{611}.

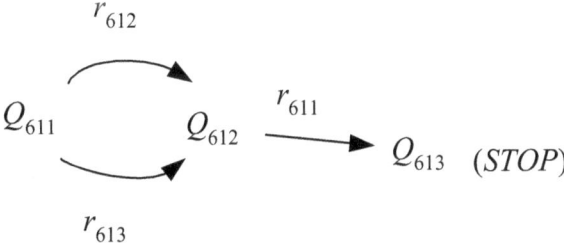

In the transition graph of the Q_{611}'s generalized SBC process, processes Q_{611}, Q_{612} and Q_{613} are defined as:

$$Q_{611} \overset{\text{def}}{=\!=} r_{612} \bullet Q_{612} + r_{613} \bullet Q_{612}$$

$$Q_{612} \overset{\text{def}}{=\!=} r_{611} \bullet Q_{613}$$

$$Q_{613} \overset{\text{def}}{=\!=} STOP$$

Although P_{611} and Q_{611} are observation equivalent, in this book we can only verify that they are not observation equivalent.

Limitation on Observation Congruence Verification

The observation congruence theory in this book has some limitations. That is, two observation congruent processes may be verified to be not observation congruent. We will demonstrate a few examples here. We just kept this limitation in this book until we find the right solution.

As a first example, consider the generalized SBC process P_{621} is defined as "$(r_{621} \bullet STOP)$" and the r_{621} (channel-based called action under a tautology condition) prefix is defined as "$<C_{621}$, CD, $k_{621}>$".

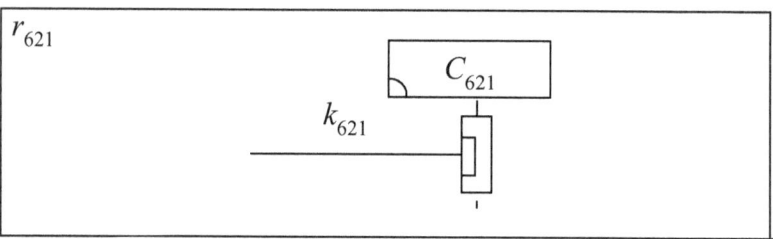

The following transition graph shows the semantics of process P_{621}.

$$P_{621} \quad r_{621}$$
$$\longrightarrow$$
$$P_{622} \quad (STOP)$$

In the transition graph of the P_{621}'s generalized SBC process, processes P_{621} and P_{622} are defined as:

$$P_{621} \stackrel{\text{def}}{=\!=} r_{621} \bullet P_{622}$$

$$P_{622} \stackrel{\text{def}}{=\!=} STOP$$

Also consider the generalized SBC process Q_{621} is defined as "$r_{622} \bullet STOP + r_{623} \bullet STOP$" and the r_{622} (channel-based called action under a certain (tautology excluded) condition) prefix is defined as "**if** (x < 300) **then** $<C_{621}$, CD, $k_{621}>$" and the r_{623} (channel-based called action under a certain (tautology excluded) condition) prefix is defined as "**if** (300 <= x) **then** $<C_{621}$, CD, $k_{621}>$".

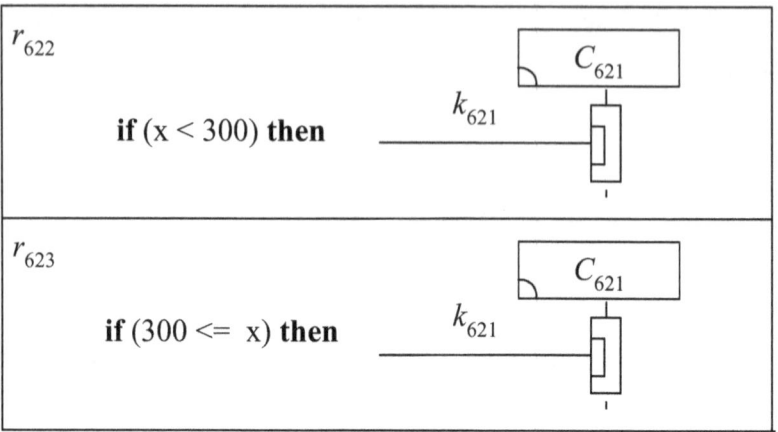

The following transition graph shows the semantics of process Q_{621}.

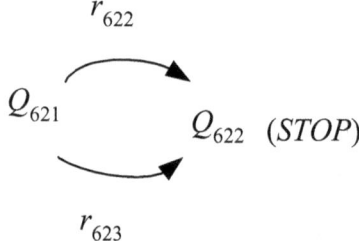

In the transition graph of the Q_{621}'s generalized SBC process, processes Q_{621} and Q_{622} are defined as:

$$Q_{621} \overset{\text{def}}{=\!=} r_{622} \bullet Q_{622} + r_{623} \bullet Q_{622}$$

$$Q_{622} \overset{\text{def}}{=\!=} STOP$$

Although P_{621} and Q_{621} are observation equivalent, in this book we can only verify that they are not observation equivalent.

As a second example, consider the generalized SBC process P_{631} is defined as "$r_{631} \bullet STOP + r_{632} \bullet STOP$" and the r_{631} (channel-based called action under a certain (tautology excluded) condition) prefix is defined as "**if** (y < 400) **then** $<C_{631},$ CD, $k_{631}>$" and the r_{632} (channel-based called action under a certain (tautology excluded) condition) prefix is defined as "**if** (400 <= y) **then** $<C_{631},$ CD, $k_{631}>$".

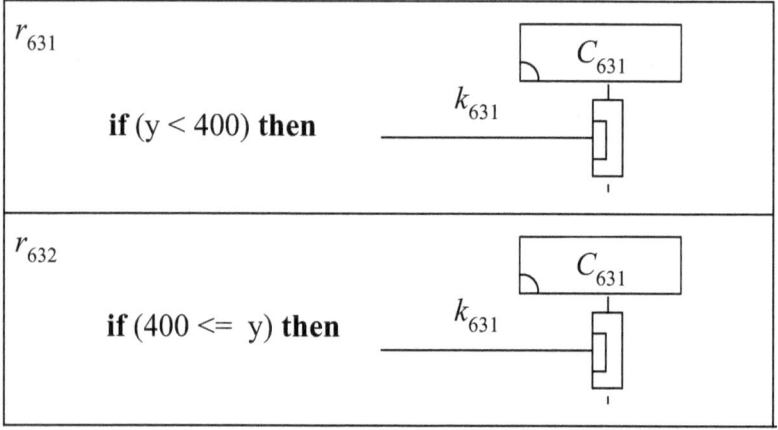

The following transition graph shows the semantics of process P_{631}.

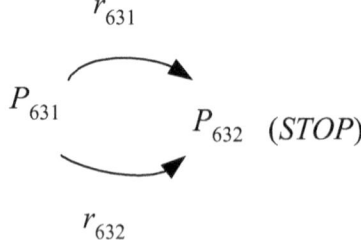

In the transition graph of the P_{631}'s generalized SBC process, processes P_{631} and P_{632} are defined as:

$$P_{631} \overset{\text{def}}{=\!=\!=} r_{631} \bullet P_{632} + r_{632} \bullet P_{632}$$

$$P_{632} \overset{\text{def}}{=\!=\!=} STOP$$

Also consider the generalized SBC process Q_{631} is defined as "$r_{633} \bullet STOP + r_{634} \bullet STOP$" and the r_{633} (channel-based called action under a certain (tautology excluded) condition) prefix is defined as "**if** $(z < 500)$ **then** $<C_{631}, CD, k_{631}>$" and the r_{634} (channel-based called action under a certain (tautology excluded) condition) prefix is defined as "**if** $(500 <= z)$ **then** $<C_{631}, CD, k_{631}>$".

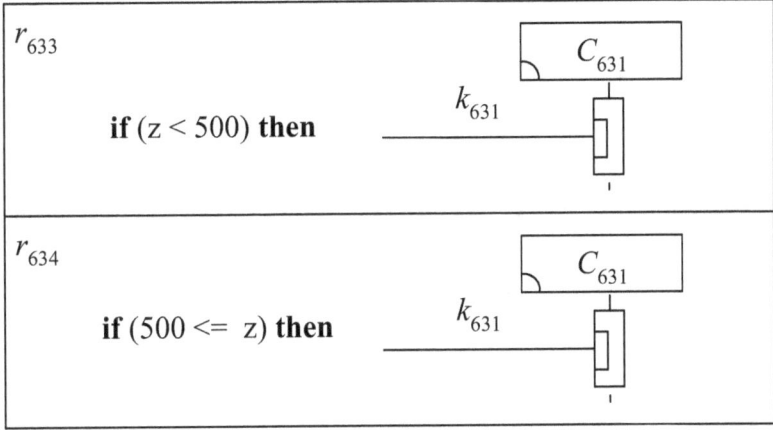

The following transition graph shows the semantics of process Q_{631}.

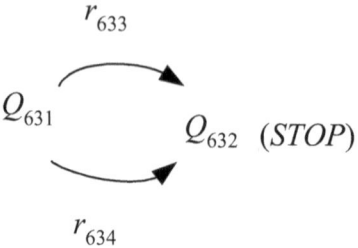

In the transition graph of the Q_{631}'s generalized SBC process, processes Q_{631} and Q_{632} are defined as:

$$Q_{631} \overset{\text{def}}{=\joinrel=} r_{633} \bullet Q_{632} + r_{634} \bullet Q_{632}$$

$$Q_{632} \overset{\text{def}}{=\joinrel=} STOP$$

Although P_{631} and Q_{631} are observation equivalent, in this book we can only verify that they are not observation equivalent.

BIBLIOGRAPHY

[Berg87] Bergstra, J. A. et al., "ACPτ: A Universal Axiom System for Process Specification," *CWI Quarterly* 15, 1987, pp. 3-23.

[Burd10] Burd, S. D., *Systems Architecture*, 6th Edition, Cengage Learning, 2010.

[Chao14a] Chao, W. S., *Systems Thingking 2.0: Architectural Thinking Using the SBC Architecture Description Language*, CreateSpace Independent Publishing Platform, 2014.

[Chao14b] Chao, W. S., *General Systems Theory 2.0: General Architectural Theory Using the SBC Architecture*, CreateSpace Independent Publishing Platform, 2014.

[Chao14c] Chao, W. S., *Systems Modeling and Architecting: Structure-Behavior Coalescence for Systems Architecture*, CreateSpace Independent Publishing Platform, 2014.

[Chao15a] Chao, W. S., *Theoretical Foundations of Structure-Behavior Coalescence*, CreateSpace Independent Publishing Platform, 2015.

[Chao15b] Chao, W. S., *A Process Algebra For Systems Architecture: The Structure-Behavior Coalescence Approach*, CreateSpace Independent Publishing Platform, 2015.

[Chao15c] Chao, W. S., *An Observation Congruence Model For Systems Architecture: The Structure-Behavior Coalescence Approach*, CreateSpace Independent Publishing Platform, 2015.

[Chao15d] Chao, W. S., *Variants of SBC Process Algebra: The Structure-Behavior Coalescence Approach*, CreateSpace Independent Publishing Platform, 2015.

[Chao15e] Chao, W. S., *Single-Queue SBC Process Algebra For Systems Architecture: The Structure-Behavior Coalescence Approach*, CreateSpace Independent Publishing Platform, 2015.

[Chao15f] Chao, W. S., *Multi-Queue SBC Process Algebra For*

Systems Architecture: The Structure-Behavior Coalescence Approach, CreateSpace Independent Publishing Platform, 2015.

[Chao15g] Chao, W. S., *Single-Queue SBC Observation Congruence Model For Systems Architecture: The Structure-Behavior Coalescence Approach*, CreateSpace Independent Publishing Platform, 2015.

[Chao15h] Chao, W. S., *Multi-Queue SBC Observation Congruence Model For Systems Architecture: The Structure-Behavior Coalescence Approach*, CreateSpace Independent Publishing Platform, 2015.

[Chao16a] Chao, W. S., *System: Contemporary Concept, Definition, and Language*, CreateSpace Independent Publishing Platform, 2016.

[Chao16b] Chao, W. S., *Systems Architecture of Electronic Toll Collection Cloud Applications and Services IoT System*, CreateSpace Independent Publishing Platform, 2016.

[Chao16c] Chao, W. S., *Systems Architecture of Smart Healthcare Cloud Applications and Services IoT System*, CreateSpace

Independent Publishing Platform, 2016.

[Chao16d] Chao, W. S., *Systems Architecture of Smart Healthcare Cloud Applications and Services IoT System*, CreateSpace Independent Publishing Platform, 2016.

[Chao16e] Chao, W. S., *Systems Architecture of Ridesharing Sharing Economy Cloud Applications and Services IoT System*, CreateSpace Independent Publishing Platform, 2016.

[Chao16f] Chao, W. S., *Systems Architecture of Handy Helper Sharing Economy Cloud Applications and Services IoT System*, CreateSpace Independent Publishing Platform, 2016.

[Chao16g] Chao, W. S., *Systems Architecture of Vacation Rental Cleaning Sharing Economy Cloud Applications and Services IoT System*, CreateSpace Independent Publishing Platform, 2016.

[Hoar85] Hoare, C. A. R., *Communicating Sequential Processes*, Prentice-Hall, 1985.

[Maie09] Maier, M. W., *The Art of Systems Architecting*, 3rd

Edition, CRC Press, 2009.

[Miln89] Milner, R., *Communication and Concurrency*, Prentice-Hall, 1989.

[Miln99] Milner, R., *Communicating and Mobile Systems: the π-Calculus*, 1st Edition, Cambridge University Press, 1999.

INDEX

www.ingramcontent.com/pod-product-compliance
Lightning Source LLC
Chambersburg PA
CBHW051635170526
45167CB00001B/206